Technology, Innovation, and Southern Industrialization

NEW CURRENTS IN THE HISTORY OF SOUTHERN ECONOMY AND SOCIETY

• This series, New Currents in the History of Southern Economy and Society, puts the history of the American South in an important new light. It demonstrates the complexity of the southern economy across time and place, and its profound impact on the varieties of people who have inhabited the region. It conveys southern economic development as consistently diverse and dynamic. The series stresses the importance of comparative and transatlantic perspectives, and strives to reintegrate southern history in relation to the rest of the nation and the globe. The series is purposefully organized around central themes rather than chronology to highlight new ways to conceptualize economic change and industrial growth.

The series originated grew out of the work of the Southern Industrialization Project (SIP), a professional organization that seeks to foster a greater understanding of the history and culture of industrialization in the American South. Each year SIP meets to discuss scholarly papers organized around key themes. This series represents SIP's commitment to developing these conference themes into substantive scholarly volumes that further promote research in southern economic and social history. In addition, SIP maintains a discussion list on H-Net of more than one hundred academic and public historians with research interests that encompass many industries, eras, and geographic locations.

Technology, Innovation, and Southern Industrialization

From the Antebellum Era to the Computer Age

Edited by

SUSANNA DELFINO AND MICHELE GILLESPIE

University of Missouri Press
Columbia and London

Copyright © 2008 by
The Curators of the University of Missouri
University of Missouri Press, Columbia, Missouri 65201
Printed and bound in the United States of America
All rights reserved
5 4 3 2 1   12 11 10 09 08

Library of Congress Cataloging-in-Publication Data

Technology, innovation, and Southern industrialization : from the antebellum era to the computer age / edited by Susanna Delfino and Michele Gillespie.

   p. cm. (New currents in the history of Southern economy and society)

  Summary: "Essays consider the role of innovative technologies in industries across the South, including steamboats and shipping in the lower Mississippi valley; textile manufacturing in Georgia, Arkansas, and South Carolina; coal mining in Virginia; sugar planting and processing in Louisiana; the electrification of the Tennessee valley; and telemedicine in contemporary Arizona"—Provided by publisher.

  Includes index.

  ISBN 978-0-8262-1795-0 (alk. paper)

  1. Industries—Technological innovations—Southern States. 2. Industrialization—Southern States. I. Delfino, Susanna, 1949– II. Gillespie, Michele. III. Series

  HC107.A13T34 2008

  338.0975—dc22   2008012922

♾™ This paper meets the requirements of the
American National Standard for Permanence of Paper
for Printed Library Materials, Z39.48, 1984.

Designer: Stephanie Foley
Typesetter: FoleyDesign
Printer and binder: Integrated Book Technologies, Inc.
Typeface: ITC Century and Palatino

Contents

v

Foreword

Industry in the Old South
Polemics and Politics

GAVIN WRIGHT

THE DEBATE OVER ANTEBELLUM SOUTHERN INDUSTRY HAS BEEN going on since antebellum times. In his famous antislavery polemic published in 1857, Hinton Helper complained:

> We [southerners] are all constantly buying, and selling, and wearing, and using Northern merchandise, at a double expense to both ourselves and our neighbors. If we but look at ourselves attentively, we shall find that we are all clothed *cap a pie* in Northern habilaments. Our hats, our cravats, our coats, our vests, our pants, our gloves, our boots, our shoes, our under-garments—all come from the North; whence, too, Southern ladies procure all their bonnets, plumes, and flowers; dresses, shawls, and scarfs; frills, ribbons, and ruffles; cuffs, capes and collars.[1]

Yet James Henry Hammond, South Carolina politician and planter, declared in 1849: "Already the South, through the almost unnoticed enterprise of a few of its citizens, more than supplies her own consumption of coarse cotton, and ships both yarn and cloth, with fair profit, to Northern markets . . . we have driven them from our markets and have already commenced the contest with them for their own."[2] In the 1850s, Athens, Georgia, was described as "like a

---

1. Hinton Rowan Helper, *The Impending Crisis of the South* (1857, reprint, Cambridge: Belknap Press of Harvard University Press, 1968), 356-57.
2. James Henry Hammond, "An Address Delivered before the South Carolina Institute at Its First Annual Fair, on the 20th November, 1849," The Making of the

northern manufacturing town"; Columbus, Georgia, was known as "the Lowell of Georgia"; and J. D. B. DeBow referred to Georgia itself as the "Empire State of the South."[3] Apprehensive statements about competition from southern manufacturers were common in the northern trade press.[4]

All of these observations were of course influenced by the economic and political motivations of the times: hostility to or support for slavery, industrial promotion, and politically motivated regional caricatures. Nor did objectivity improve in the postbellum era, when New South propagandists never tired of the rhetorical formula that the abolition of slavery had emancipated southern whites as well as blacks. There were countless variations on the theme articulated by journalist Hoke Smith: "Had it not been for the institution of slavery . . . the South, with natural resources in its favor in 1860, would have been the greatest manufacturing and mining, as well as agricultural, section in the Union."[5]

Unfortunately, the inclination to hype or denigrate antebellum southern industry has carried over into modern scholarship. Critics note that the South accounted for barely 10 percent of the nation's manufacturing output in 1850, and this share hardly changed in the subsequent decade. The agricultural Midwest surpassed the whole of the South in manufacturing output, the gap widening markedly during the 1850s.[6] Defenders respond that southern industry was "backward" only in comparison with that of the U.S. North, the emerging industrial leader of the world. By international standards, indicators of economic development in the South were quite respectable, far ahead of slave societies elsewhere, such as Brazil.[7] Regional differences that did exist do not necessarily reflect intrinsic weaknesses in culture and institutions but may be readily

---

Modern World: The Goldsmith Kress Library of Economic Literature (Farmington Hills, MI: Thomson Gale, 2006), 17–18.

3. All quoted in Jonathan Daniel Wells, *The Origins of the Southern Middle Class 1800–1861* (Chapel Hill: University of North Carolina Press, 2004), 24-25, 168.

4. *Ibid.*, 208–15.

5. Quoted in Joseph J. Persky, *The Burden of Dependency: Colonial Themes in Southern Economic Thought* (Baltimore: Johns Hopkins University Press, 1992), 99.

6. Fred Bateman and Thomas Weiss, *A Deplorable Scarcity: The Failure of Industrialization in the Slave Economy* (Chapel Hill: University of North Carolina Press, 1981), tables 1-2.

7. Richard Graham, "Slavery and Economic Development: Brazil and the United States South in the Nineteenth Century," *Comparative Studies in Society and History* 23 (1981): 620-55.

explained by basic economic concepts: comparative advantage, population density, climate, and the limited "linkage" opportunities provided by southern as compared to northern staple crops. Since all of the preceding statements are true, the dialogue may continue indefinitely, with little prospect of resolution or redefinition.

Surely it is time to move beyond this "fully-only" interchange. An alternative is to descend from the heights of macro generalizations into the real historical worlds of entrepreneurs, developers, and machinists trying to make a go of it in the antebellum South. What were their social and technological backgrounds, their hopes and plans, their sources of support, and their politics? What problems did they encounter in production, in labor relations, and in marketing? How did the competitive atmosphere compare with that in the North? Such an approach may be labor-intensive and lacks the geographic range and statistical authority of census-based methods. But in exchange for their limitations, case studies offer the promise of new insights and reformulations prompted by historical experience on the ground rather than by theoretical abstractions.

The essays in this volume advance this agenda admirably. Michele Gillespie introduces the colorful character Henry Merrell, a mechanic and mill manager who left Oneida, New York, in 1839 to seek his fortune in the South. After working his way up in management, Merrell founded three cotton mills in Georgia in the 1840s. Despite Merrell's skills and work ethic, however, these ventures were financial disasters. Identifying specific reasons for the failure of specific enterprises is always difficult. But from Gillespie's account, it is evident that Merrell had difficulties with the quality of the regional labor, with capital constraints, and with long-distance interactions with machinery suppliers. These features of the antebellum setting are symptomatic of early industrialization, when a new region enters into competition with a more established industrial center. They invite comparison with postbellum developments in the same geographic area, when the same challenges were much more readily overcome.

But individual cases do not always conform to analytical generalities. While many Americans headed for the frontier to escape debts, the ethical Henry Merrell remained in Georgia for more then five years, working to pay off his debts. In 1856 he finally achieved industrial success, in an unlikely, remote location in Arkansas! Such twists of history are chastening for economists, as they should be.

Because Arkansas never became a major textile center, however, it is reasonable to assume that Merrell's formula for success did not lend itself to ready extension or replication.

Sean Patrick Adams presents a different comparison in his account of frustrated development in the Richmond, Virginia, coal basin. Rather than a new regional entrant up against an established northern industry, Richmond coal was seen by Hamilton and Coxe as a promising potential growth center for the new nation. But on one count after another—labor, capital, transportation, and technology transfer—Virginia colliers encountered disappointment and failure. Dealings with outside machinery suppliers were particularly severe, highlighted in dramatic fashion by Oliver Evans's refusal to send workmen to assist with steam-power technology, owing to their "prejudicial idea of the customs of your country." One could hardly ask for clearer evidence that regional caricatures go back to the very beginning of national history. But outright northern noncooperation was evidently not the primary reason for the stagnation of the Richmond coal industry.

Adams assigns ultimate causal significance to politics, noting that technology transfer is an ongoing process requiring social and political support. He contrasts Virginia with Pennsylvania, where rapid development of the anthracite region was actively fostered through state-sponsored canals and corporate charters.

Hammond might have said something similar about the textiles industry. In the same 1849 address quoted above, in which he boasted of the region's emerging prowess in textiles, Hammond lamented that "the political influence of the manufacturers of the South is nothing. It cannot send a single representative to Congress, perhaps not even to a State legislature."[8] The emphasis on political support seems to return us to an earlier era in the historiography of the South. Since the advent of cliometrics, a primary objective of economic history has been to explain the course of economic events in terms of economic variables—costs, markets, technology, and competition—escaping the Beardian fallacy that economic outcomes reflect the collective choice of politically dominant classes. Have we now come full circle?

I think not. This is not the old political history, in which the analytical challenges of explaining collective choice were blissfully

8. Hammond, "An Address," 17.

waived aside. In American history, "political support for economic development" rarely if ever took the form of central planning and direction. Rather, governments responded to the pluralistic pushes and pulls of contending interest groups, typically following the path of least resistance by giving each major group its most-favored benefit, unless, of course, this first-choice favor was directly opposed by an even more powerful interest group. This is the sense in which the politics of the slave regime posed a problem with which fledgling southern industrialists had to deal. In the North, the pro-growth program unified a coalition that included farmers, developers, canal and railroad builders, and merchants. The same actors were present in the South as well, and they were often successful in gaining legislative support.[9] But the first political priority of slaveholders was maintaining the security of slavery. Thus the essays in this volume contribute to a broader transformation in the field of economic history as appreciation for the interactive character of political and economic change has grown in recent years.

Lack of political clout was hardly a problem for the Louisiana sugar planters analyzed by Richard Follett. As he shows, these wealthy slave owners were energetic and acquisitive, making effective use of the state's advanced banking system to finance a sophisticated steam-powered sugar technology. Apparently, with sufficient deployment of wealth and influence, technology transfer was not an insuperable problem for the antebellum South. Also key to the sugar planters' prosperity was their success in lobbying the federal government for tariff protection, an essential prerequisite for the entire operation.

Yet despite the extraordinary riches accumulated under these arrangements, Follett indicts the sugar planters for thinking only of themselves, failing to foster (in Olmstead's words) an *"atmosphere* of progress and improvement" as in the northern states. How ironic that the acquisitive, individualist North should be commended for its relative civic-mindedness! An interpretation based on coalition-formation and political priorities would seem more promising than one based on constricted vision. On reflection, perhaps it is understandable that sugar planters might have had difficulty picturing

---

9. Most recently, see Tom Downey, *Planting a Capitalist South: Masters, Merchants, and Manufacturers in the Southern Interior, 1790–1860* (Baton Rouge: Louisiana State University Press, 2006).

just what a civic-minded, progressive community-development program would have looked like in a society composed of vast slave plantations.

How the steamboat fits into the antebellum regional picture is another challenge, and it is posed by Robert Gudmestad's innovative essay. Here, too, technology transfer seems not to have been a problem, and Gudmestad notes that the early steamboat business featured "a high degree of cross-sectional cooperation." Indeed, he might have gone further by calling attention to the rapid progress in steamboat technology during the antebellum era. The full power of the emerging American technological network was deployed toward the objective of adapting steamboat design for the shallow-water conditions of internal waterways, in some ways the country's most significant technical achievement during that era.[10] Although the original innovators were easterners (including the ubiquitous Oliver Evans), the adaptation and improvement process had eager participants from both slave and free states.

Gudmestad suggests, however, that this form of technological progress was acceptable in the South because it passed a political test: steamboats "accelerated the development of the plantation economy which ensured the spread and maturation of the slave system. . . . [T]hey reinforced the hegemony of the plantation system." Although advances in river transportation undoubtedly did strengthen the plantation economy, one must remember that the uses and benefits of the system were national rather than narrowly regional, encompassing such economic interests as lead mining, the fur trade, grains, and, of course, passengers. Capital requirements for individual steamboat enterprises may have been modest, but expenditures for snag-clearing and other river improvements constituted one of the largest federal infrastructure investments of the antebellum era. Its constitutionality and regional equity were fiercely debated across the entire antebellum era, creating highly unlikely political bedfellows in the process.[11] Simple generalizations

10. The best account of technical developments is still Louis Hunter, *Steamboats on the Western Rivers* (Cambridge: Harvard University Press, 1949), chap. 2.

11. Paul F. Paskoff, "Hazard Removal on the Western Rivers as a Problem of Public Policy, 1821–1860," *Louisiana History* 40 (summer 1999): 261-82. For federal expenditures, see Susan B. Carter et al., *Historical Statistics of the United States: Millennial Edition* (New York: Cambridge University Press, 2006), tables Df13–16, vol. 4, p. 780.

about connections between steamboat politics and slavery should be resisted: If slavery had been abolished in 1790 and the South settled by ambitious family cotton farmers on the midwestern model (truly the "path not taken" in American history), would the region have been any less enthusiastic about the steamboat?

The volume rounds out with three essays on three different phases of post–Civil War southern economic history. Pamela Edwards shows the roles of diverse national networks (financial, business, and technical) in promoting southern industrialization from 1890 to 1925. Stephen Wallace Taylor recounts the changing priorities and rhetoric of the Tennessee Valley Authority from its inception in the 1930s through the energy crisis of the 1970s. And Yoneyuki Sugita describes the catch-up strategies pursued by the state of Arizona in recent years, as a late entrant into interstate competition for business. Even for an intellectual project with its heart in the antebellum era, attention to postbellum counterparts is essential. They call attention to the very different demands and success criteria facing entrepreneurs and industrialists in later periods, when the flow of national and international currents was going their way, as contrasted to the early southern businessmen who pursued uphill struggles under the slavery regime.

Taken as a whole, the volume marks a distinct advance in the study of southern industrialization and a hopeful step forward in building a truly interdisciplinary economic history. Both economics and history have much to gain from breaking down the longstanding walls of segregation between them. The essays that follow can be cheerfully recommended to readers on both sides of the aisle.

Technology, Innovation, and Southern Industrialization

Introduction

SUSANNA DELFINO and
MICHELE GILLESPIE

FOLLOWING THE PUBLICATION OF ALEXANDER HAMILTON'S *Report on Manufactures* in 1791, Virginia slaveholder and statesman Thomas Jefferson reluctantly acknowledged that industry might be as important as agriculture and commerce to the new republic's future. Early manufacturers, merchants, and even many agriculturalists had little problem embracing that concept by comparison and, in fact, bandied about terms like *industriousness, manufacturing,* and the *useful arts* to describe all three modes of production, which they considered tightly linked. The arrival of the textile industry, with its rapid mechanization, along with the emergence of other important industries such as flour-milling, iron making, and lumber milling brought on a degree of economic transformation that can only be described as a revolution, thrust industry into the forefront of agriculture and commerce, and cultivated a new culture in which mechanical knowledge became a kind of commodity in and of itself. In short order, the search for techniques, devices, and machines to facilitate industrialization was being fueled by eager capitalists through patents, public funds, technical journals and books, professional societies, and especially the importation (albeit not always legal) of skilled workers and machinery from Britain.[1]

Thus the emergence of the market economy in the Early Republic launched a kind of larger cultural understanding of the value of

---

1. John Kasson, *Civilizing the Machine: Technology and Republican Values in America* (New York: Grossman, 1976).

inventive genius that remains characteristically American. Borrowed in part from eighteenth-century Enlightenment intellectuals, this emerging emphasis on the importance of innovation and human ingenuity has developed into a kind of yardstick for measuring a society's advancement.[2] The degree to which a society has harnessed innovation (the act of inventing something new) to develop fresh technologies (the applications of those innovations) is considered central to sustained economic growth and the improvement of human civilization. Perhaps not surprisingly, just as Western societies long privileged their "innovativeness" over their assumptions about the absence of "innovativeness" in most other non-Western countries, the northeastern United States privileged its "innovativeness," as measured by the speed with which it industrialized and by the sophistication of its mechanical inventiveness, over other regions of the country, especially the South. Because the South remained predominantly agricultural with its slaveholding economy, attention focused on the northeast, with its shockingly "new" manufacturing society and free labor economy. The assumption that "new" processes could only happen in structurally transforming places prevented even historians, who should have known better, from looking for innovations and new technologies taking place in the South. Fortunately, as scholars in the last half of the twentieth century began to ask better questions about the role of technology and innovation in the processes of industrialization that took place in different parts of the United States, a few historians recognized the cultural blinders that had prevented them from seeking out innovations and new technologies in other places and regions, and the American South in particular. Thus, in formulating this volume, it was our goal as editors to bring together a group of essays that pressed historians *not* to presume that southerners eschewed innovations or that the South lacked new technologies long after other regions of the United States and other Western countries had embraced them. Instead, these essays explain the multiple contexts in which innovation and the adoption of new technologies could occur in the South and the technologies' varying degrees of success, in specific time periods, industries, and places.

Technological advances and inventiveness in devising ever-better methods of production traditionally have been considered

2. Franklin L. Baumer, *Modern European Thought: Continuity and Change in Ideas, 1600–1950* (New York: Macmillan, 1977).

indispensable elements in fostering a process of industrialization as well as crucial indicators of the "genuineness" of its onset. The issue of technology, however, has acquired a more nuanced meaning ever since the notion of the existence of a multiplicity of ways to industrialization has prevailed over the previous approach, which had considered the experience of Great Britain, the "first comer," as a model against which to gauge the industrial development of all other nations.

Between the late 1950s and the early 1960s, Alexander Gerschenkron argued that, although eighteenth- and nineteenth-century nations were understandably very jealous of their inventions, there was nothing diminishing—in terms of the "genuineness" of the process of industrial development—for one country to adopt imported technology.[3] In other words, once new machines or innovative methods of production were devised in one country, they became the patrimony of human society at large. Expanding on this concept, while circumscribing their arguments to the British-and-American experience, H. J. Habakkuk and David Jeremy later showed that the industrialization of the United States took place thanks to the transfer of British technology, which was "adapted" to meet the specific characteristics of the production factors present on the American soil.[4]

The concept of "adaptation"—rather than mere imitation—hints at the issue of innovation as a crucial phenomenon that, supposedly, can only spring up in a cultural milieu conducive to industrialization. As Habakkuk, Jeremy, and others explained, out of consideration of the differential between the cost of investment in new technology and the rates of return expected from the sale of their products, American manufacturers found it more convenient to exploit the greater abundance of wood to make implements, tools, and machines than was true with the British. For the same reason, until the 1840s, charcoal remained the most extensively used fuel to smelt iron in the United States, although the British had

3. Alexander Gerschenkron, *Economic Backwardness in Historical Perspective* (Cambridge: Belknap Press of Harvard University Press, 1962), chap. 2.

4. H. J. Habakkuk, *American and British Technology in the Nineteenth Century* (Cambridge: Cambridge University Press, 1962); David Jeremy, "Innovation in American Textile Technology during the Early Nineteenth Century," *Technology and Culture* 14 (1973): 40-76; David Jeremy, *Transatlantic Industrial Revolution: The Diffusion of Textile Technology Between Britain and America, 1790–1830s* (Cambridge: MIT Press, 1981).

turned to coke since the eighteenth century.[5] In other words, the kind of technology used in one country cannot be taken as an unequivocal indicator of the stage of its industrial development. Rather, it is better understood as the expression of rational choices based on consideration of the comparative advantage offered by the exploitation of the endowment factors specific to it. In this light, the decision to use the newest technology is determined by the calcula-tion of the differential between the cost of its application and the profits expected by the anticipation of a market expansion.

In the early 1980s, economic historian Thomas Cochran offered a more sophisticated picture of the spreading of technology in the north-Atlantic world. Instead of a simple "transfer" from Britain to the United States, he saw an "interchange" between the two coun-tries starting at least from the 1800s. As an example, he mentioned the case of nail-cutting machinery, an American advance by Jacob Perkins, which slowly reverberated back to Great Britain.[6] Further-more, Cochran offered a useful insight on Americans' perceptions of the significance of their own industrial development in relation to the Western world, generally, and more specifically, to the North/South relationship within the country. Noting that the rate of growth in the United States during the antebellum period was by no means exceptional as compared with that of other countries of the Western world, and that industrialization was originally a regional phenomenon, in the United States and elsewhere, he remarked that, an ingrained belief in the uniqueness and superior-ity of their society combined with religious faith led northeastern Americans to perceive their region as the most rapidly growing not only within the country, but also in the Western world.[7]

The level of demand is a pivotal factor in stimulating both the invention of new technology and the extensive adoption of an

5. Habakkuk, *American and British Technology,* 48-52; Jeremy, "Innovation in American Textile Technology"; Peter Temin, "Steam and Waterpower in the Early Nineteenth Century," and "A New Look at Hunter's Hypothesis about the Ante Bellum Iron Industry," both in *The Reinterpretation of American Economic History,* ed. Robert W. Fogel and Stanley L. Engerman (New York: Harper & Row, 1971), 129, 117; Paul W. Strassman, *Risk and Technological Innovation: American Manufacturing Methods during the Nineteenth Century* (Ithaca: Cornell University Press, 1959).

6. Peter Temin, *Iron and Steel in Nineteenth Century America* (Cambridge: Cambridge University Press, 1964), 17; Thomas C. Cochran, *Frontiers of Change: Early Industrialism in America* (New York: Oxford University Press, 1981), 17, 50-52.

7. Thomas Cochran, "The Paradox of American Economic Growth," *Journal of American History* 61 (1975): 925-41.

already existing one. In this regard, the timing and pace of the mechanization of American agriculture constitutes a case in point. If, on the one hand, the invention of the cotton gin in 1793 by Massachusetts-born and Georgia-transplanted Eli Whitney was fostered by the soaring demand for raw cotton to supply the burgeoning British textile industry, the extensive adoption of agrarian technology in the Midwest in the early 1850s occurred thanks to the rising demand for American grains during the Crimean War in Europe. Even in the northeastern United States, however, where advanced technology applied to agriculture had been adopted since the early decades of the century, probably less than one-fourth of farmers made use of modern equipment as late as 1850. On the other hand, the reaping machines invented by Obed Hussey and Cyrus McCormick in the 1830s were not widely utilized in the Midwest until the 1850s. The same was true for threshers and mowers, as well as for steel breakers and plows. Moreover, as late as the 1880s, mechanical innovation in the Midwest was still the preserve of larger-than-average farms where owners had money to invest.[8]

In the South during the 1830s, as reported by visitors such as New Yorker Frederick L. Olmsted and the Englishman Henry Barnard, the use of the most advanced agrarian technology was largely restricted to wealthy planters. In the following decades though, the South, like other American regions, witnessed the adoption of most modern farm technology. As Fogel and Engerman noted, in 1860 "expenditures on farming implements and machinery per improved acre were 25 per cent higher in the seven leading cotton states than they were for the nation as a whole."[9] Ultimately, the extensive adoption of new technology in farming depended on regional differences in the quality of soil and in crops raised. For instance, the cast iron plows used in New England and the Mid-Atlantic states in the 1830s were not suitable to cut the tough prairie

8. Peter Temin, *Causal Factors in American Economic Growth in the Nineteenth Century* (London: Macmillan, 1975), 33; Paul A. David, "The Mechanization of Reaping in the Antebellum Midwest," in *The Reinterpretation of American Economic History*, 217, 225; Roy Bainer, "Science and Technology in Western Agriculture," *Agricultural History* 49 (1975): 55-56; Reynold M. Wik, "Some Interpretations of the Mechanization of Agriculture in the Far West," *Agricultural History* 49 (1975): 76.

9. Cochran, *Frontiers of Change*, 87; Frederick L. Olmsted, *A Journey in the Seaboard Slave States, with Remarks on Their Economy* (New York: Dix & Edwards, 1856), 669; Robert W. Fogel and Stanley L. Engerman, *Time on the Cross: The Economics of American Negro Slavery* (Boston: Little, Brown, 1974), 255, 265.

soil of the West. By the same token, the mechanical reaper was not suitable for corn, of which the South produced half the national crop in 1850.[10]

According to the *Census of Agriculture*, in 1860 the value of farm implements and machinery ascertained for most southern states was higher than that showed by the majority of either the western or the northeastern ones. Louisiana ranked third, after Pennsylvania and New York, in this regard.[11] The reason for this was that Louisiana was a major user of steam engines, which generally constituted a large share of the value of the agricultural machinery found in the South. There, they were extensively used in the processing of sugar, lumber, rice and flour, as well as of cotton. According to the congressional *Report on Steam Engines* of 1838, 28 percent of the steam engines identified in the United States were located in the South. Peter Temin found that approximately one-third of the steam engines used in the South of which the origin is known, mainly the high-pressure type, were built there; however, the report also indicated a large number of steam engines the origins of which were unknown, and they could easily have been built were in the South.[12]

Demand remained central for stimulating the adoption of advanced technology in the manufacturing sector, too. In fact, it greatly affected the mechanization of cotton weaving in New England, which occurred between 1808 and 1815, thanks to the protection given by the interruption of foreign trade. Between 1790 and 1814, carding machinery and water-powered spinning jennies, mules, or frames were introduced in America. In 1814, the first cam-driven loom, an American improvement on the power loom designed by Edmund Cartwright in England, began to operate at Lowell, Massachusetts. By 1825, American cotton textile manufacturing had become completely mechanized.[13] But if the growing domestic demand for goods was of paramount importance in the textile industry to stimulate the adoption of devices apt to produce

10. Temin, *Causal Factors*, 35-39; Fogel and Engerman, *Time on the Cross*, 277.

11. U.S. Bureau of the Census, *Agriculture of the United States, 1860* (Washington, DC: Government Printing Office, 1861), cxxvi-vii.

12. Peter Temin, *Steam and Waterpower* (Cambridge: MIT Press, 1964, 229-30; Susanna Delfino, *Yankees del Sud: sviluppo economico e trasformazioni sociali nel Sud degli Stati Uniti* (Milan: Franco Angeli, 1987), 72-74.

13. Robert B. Zevin, "The Growth of Cotton Textile Production after 1815," in *The Reinterpretation of American Economic History*, 141; Cochran, *Frontiers of Change*, 62-63.

more and faster at lower costs, the history of American iron making suggests a different path. According to some authors, the wavering tariff policy of the United States during the later part of the antebellum period did not so much protect American iron as, rather, prolong the use of charcoal and retard the adoption of improved techniques. In the 1840s, 58 percent of the nation's furnaces still used charcoal, and in 1858 this proportion had decreased by only 14 percentage points.[14] This was because, in the face of formidable competition by British products, the continued use of charcoal represented a plus for American ironmasters, in that it allowed for the production of better-quality iron, although at higher costs. Unable to compete with imported British iron, they found a market niche in their own country for applications in which high-quality iron was important, even as a movement for the emancipation from dependence on massive importation of British iron became stronger and stronger from the 1840s. Two of the leading figures in the promotion of the American iron industry during the Jacksonian era were the well known former journalist and publisher Duff Green, who had become a full-time businessman and had moved to Dalton, Georgia, by the late 1840s, and South Carolinian Franklin H. Elmore, the owner of the Nesbitt Manufacturing Company and, later, the successor of John C. Calhoun in the U.S. Congress.[15] In effect, the decline of the charcoal iron industry, from the 1840s, was due less to British competition than to the domestic technological progress connected with the use of anthracite.[16]

These examples convey how current scholarship looks at the factor of technology in the early industrialization of the United States as one of diverse regional, and sometimes subregional, impact to be

14. Lester J. Cappon, "Trends of the Southern Iron Industry under the Plantation System," *Journal of Economic and Business History* 2 (1929–1930): 376-77; Robert W. Fogel and Stanley L. Engerman, "A Model for the Explanation of Industrial Expansion during the Nineteenth Century," in *The Reinterpretation of American Economic History,* 154; Frank Taussig, *The Tariff History of the United States* (New York: G. P. Putnam's Sons, 1931).

15. Duff Green, "What Will Be the Consequences of Continuing to Import Foreign Iron for Our Railroads?" Undated Business Papers, Duff Green Papers, Southern Historical Collection, University of North Carolina, Chapel Hill; Susanna Delfino, "'To Maintain the Civil Rights of the People': The Tribulations of Duff Green, Iron Manufacturer in Civil War East Tennessee," *Journal of East Tennessee History* 72 (2000): 51-52; Temin, *Iron and Steel,* 17-38.

16. Fogel and Engerman, "A Model for the Explanation," 159; Paul F. Paskoff, *Industrial Evolution: Organization, Structure, and Growth of the Pennsylvania Iron Industry, 1750–1860* (Baltimore: Johns Hopkins University Press, 1983).

matched carefully against the British experience. However, for all the articulation and interpretational relativism introduced by the new approaches, the study of technology in early American industrialization has mostly pertained to the northeast United States while the South has been left out of this sort of investigation, under the apparent assumption that no process of industrialization had taken place within its bounds during the antebellum period. Notwithstanding the growing number of scholarly contributions that have suggested the contrary, the issue of technology in the South has not elicited much attention. This very fact reflects the still-immature stage of studies on the southern manufacturing sector, as well as the belief that, in the antebellum South, even agriculture had not undergone a process of modernization owing to its labor force being primarily slaves.[17]

This volume, the second of the series New Currents in the History of Southern Economy and Society, begins to fill this void, introducing topics of analysis and discussion that the series editors hope will help broaden attention on this important aspect of the South's economic development during the antebellum era. According to the more traditional interpretation, the fact that southern factory owners as well as planters largely imported their machinery from the North—mainly from New Jersey, Pennsylvania, and New York—when not from England, was a sign of the region's industrial backwardness. This assumption, as we have seen, has increasingly been challenged as a truthful indicator of the onset of a process of industrialization. Southern manufacturers and entrepreneurs did, in fact, rely on the wide supply of machinery available from the Northeast and applied the most advanced techniques in their respective areas of activity. In mining, for instance, as early as 1818, the Virginian Chesterfield Company used a low-pressure steam engine to pump water out of the deepest pits according to the method then in use in Cornwall. Around 1840, the Black Heath Company, which operated in the Richmond Basin, could be considered at the national vanguard in technological progress. It could boast the deepest shaft (seven hundred feet) to be found in the whole country, and it made use of safety lamps and other equipment of the most advanced models. As reported by

17. Eugene D. Genovese, *The Political Economy of Slavery: Studies in the Economy and Society of the Slave South* (New York: Random House, 1965), part 2, chap. 6.

French minister Michel Chevalier, who visited the country in the early 1830s, even railroads, the most recent and exciting invention, could be found at the bottoms of the largest Virginia mines.[18] These and other similar examples have however been interpreted as isolated cases, more the expressions of the foolish ambitions of an elite of wealthy entrepreneurs than a signal of widespread regional modernization.

However, between the 1840s and the 1850s, when the railroad revolution swept the South, giving rise to a strong demand for both industrial and agricultural machinery and tools, foundries and machine shops such as those managed by Leach and Avery, Skate and Company, and William Anderson, respectively, in Alabama, and the Eagle Machine Foundry in Virginia and the Phoenix Ironworks in South Carolina, started to bloom in the southern states.[19] The rise of a southern machine industry contributed to making the demand for machinery from the Northeast and from Great Britain increasingly confined to the most sophisticated articles, especially in textile manufacturing. Not infrequently, wealthy entrepreneurs with great ambitions as would-be manufacturers tried to secure the most recent technological innovations for their factories. For example, in 1849, when Farish Carter, reputed to be the wealthiest planter in Georgia, was about to establish a cotton manufacture, the Coweta Falls Manufacturing Company, he relied on the advice of John Craig of Tallapoosa, an experienced technician. Craig installed in Carter's factory a "ring and traveller," a new type of frame that was used in England under the utmost secrecy but that Craig could apparently replicate.[20]

Balancing the costs involved, southerners tried as much as possible to introduce the most advanced technology in use in Great Britain in iron making as well. So, for instance, by the mid-1850s the puddling and rolling techniques for iron processing had become quite popular among Virginia, Tennessee, Georgia, and Alabama ironmasters. Toward the end of the antebellum period, there were

18. Michel Chevalier, *Society, Manners and Politics in the United States* (Boston: 1839), 84; Kathleen Bruce, *Virginia Iron Manufacture in the Slave Era* (New York: Century Company, 1931), 82-83, 95.

19. Delfino, *Yankees del Sud*, 75, 43n.

20. John Craig to Farish Carter, July 12 and 13, 1849, Farish Carter Papers, Southern Historical Collection, University of North Carolina, Chapel Hill, quoted in Delfino, *Yankees del Sud*, 75-76.

also successful attempts to switch to coke in the smelting of the mineral ore, one major example constituted by the Bluff Furnace of Chattanooga.[21]

The antebellum South, however, was not a mere imitator of technology devised elsewhere. In the late 1840s, a Kentuckian, William Kelly, was in fact the inventor of a new method to make steel, one known as Kelly's "air boiling process," which was the precursor of the Bessemer converter. Georgian Robert Findlay, the founder of the Findlay Steam Engine Manufactory in Macon, was an important innovator, if not an inventor proper. His machines, especially his steam engines, were renowned throughout the South, and one of them survived intact to World War II.[22]

These examples invite a reappraisal of the southern case during the early phases of its process of industrialization even though, as David Carlton and Peter Coclanis have argued, while the still-immature stage of industrial development of the whole United States made the antebellum South figure decently in the field of technology, the gap between North and South in this regard widened after the Civil War.[23] This was presumably due to the influence of labor costs in determining the competitiveness of the finished products. While in the Northeast the requests advanced by unionized labor forced manufacturers, especially in the field of textiles, to intensify their use of labor-saving devices to retain their market competitiveness, the low cost of labor in the South did not make that choice as compelling. Arguably, the globalizing trends that shaped both the national and international economy after the Civil War speak to a scenario completely different from that of the antebellum period; in the former, the defeated South participated in the industrial growth of the United States as a region dramatically

---

21. Cappon, "Trends of the Southern Iron Industry," 371-75.

22. William Kelly was born in Pennsylvania but moved to Kentucky as a young groom. He was the owner of Union Furnace in Kentucky and Suwanee Furnace in Tennessee. He conceived the idea of cold blast to make malleable iron from pig metal without fuel. When he heard of Bessemer's obtaining a patent for his tilting converter in 1855, he claimed prior invention and was issued a U.S. patent in 1857. Robert B. Gordon, "The "Kelly" Converter," *Technology and Culture* 33 (1992): 769-79; Cappon, "Trends of the Southern Iron Industry," 367-68; Robert S. Davis, *Cotton, Fire, and Dreams: The Robert Findlay Iron Works and Heavy Industry in Macon, Georgia, 1839–1912* (Macon, GA: Mercer University Press, 1998), chap. 3.

23. David L. Carlton and Peter Coclanis, "The Un-Inventive South? A Quantitative Look at Region and American Inventiveness," *Technology and Culture* 36 (1995): 302-26.

impoverished and impaired in its industrial capacity by the ravages of the recent conflict.[24]

The nine essays in this volume examine these critical questions. The collection crosses three centuries of southern history, from the antebellum era into the twenty-first century, and is organized chronologically. It includes most of the southern states, from the Atlantic coast states to the Gulf states and into the southern interior, but it also expands the traditional definition of "southern states" to include Arizona as part of the recent "Sunbelt South" phenomenon. Four of the essays look at four different industries in four different subregions of the antebellum South: steamboats in the lower Mississippi valley, textiles in Georgia and Arkansas, coal mining in Virginia, and the sugar-processing industry in Louisiana. The remaining essays examine the importance of technology in late-nineteenth- and early-twentieth-century South Carolina textiles, the electrification of the Tennessee valley in the mid-twentieth century, and telemedicine in late-twentieth- and early-twenty-first-century Arizona.

Robert H. Gudmestad invites us to understand the critical impact of steamboat technology west of the Appalachian Mountains on the development of the cotton kingdom and the diversification of the southern economy. Once the first northern-made and northern-capitalized steamboat traversed the Mississippi River system in 1811, southerners so quickly recognized the potential of this rivercraft that they embraced this new technology by immediately building their own. In short order, steamboats evolved into especially powerful symbols of a particular kind of southern world. Gudmestad argues that steamboats, in fact, sped up the formation of the antebellum southern order by fostering the spread of plantation culture and city growth up and down the western rivers. Southerners enthusiastically harnessed this new technology, which suited their agricultural world so well, using it to expand their slave society and making slavery more adaptable in the process.

While steamboats signify the ways in which southerners wholeheartedly embraced technological innovation and adapted it to their own social and economic needs beginning in the early nineteenth century, Sean Patrick Adams offers an example of failed

24. Carlton and Coclanis, "The Un-Inventive South?"; Robert A. Margo, "The South as an Economic Problem: Fact or Fiction?" in *The South as an American Problem,* ed. Larry J. Griffin and Don H. Doyle (Athens: University of Georgia Press, 1995), 166-80.

adaptation of technology in the coal mining industry in Virginia. Eager to cultivate a national market for coal at the end of the War of 1812, investors in the Richmond basin implemented a whole slew of sophisticated methods borrowed from the British coal mining industry, from shaft construction to ventilation and coal extraction, with limited success. But the difficulties of implementing steam engine technology, improved transport, and managing production in this particular place at this particular time prevented the colliers from realizing their ambitions. Adams argues that while the prevailing social order of agricultural slavery never thwarted their efforts outright, it did "chip away" at their competitiveness, whether in respect to the availability of labor, the passage of legislation that favored agricultural interests over theirs, or the impediments to improved transportation systems.

Technology transfers that succeeded in the antebellum South, then, were those that unambiguously facilitated the development of slave-based crop cultures. Richard Follett underscores this critical point in his analysis of technology in the antebellum Louisiana sugar industry. Louisiana planters pursued technological innovations while balancing them against the costs of sugar producing. They aggressively sought out new technologies that sped up production in an especially time-conscious industry, and "modernized" all the more by literally making their slaves work like machines. Louisiana sugar planters poured ten times more capital into implement and machinery investment than their counterparts in the cotton South, or wheat farmers in the Midwest. But sugar planters' insistence on "operating as lone producers in an increasingly competitive world market" proved to be their undoing over time. In the end, Follett accuses these planters of self-absorption and regional myopia. Too narrowly focused on their own identity as enterprising slaveholding individualists in the deep South, they remained fearful of the most expensive innovations, grew too conservative in the wake of the volatile sugar market and tariff laws, and failed to recognize the value of cooperation with one another to invest in shared refinery equipment.

Michele Gillespie invites us to look at the perspective of a transplanted northern mechanic to understand the larger problems inherent in the fledgling southern textile industry. Henry Merrell hoped to recreate in Roswell, Georgia, the thriving textile world characterized by both profitability and a Christian-based work ethic from the Northeast, which he believed benefited employer and employee alike. Although Merrell commanded the technical knowledge and access to technical communities that could both improve the quality of textile production and make it quicker, more efficient, and more reliable, investors' qualms, as well as their insistence on the primacy of the cotton market, thwarted his efforts. Too frequently, he gambled with his lack of capital resources and proved unable to take advantage of the newest and highest-quality innovations, propelling him into bankruptcy on several occasions. He recognized that only when cotton planters-turned-halfhearted-textile investors made significant and sustained outlays in capital, thought more strategically about the relationship between local and global competition, and cooperated more fully with one another by sharing technical information could the southern-based textile industry be truly profitable.

The unquestionably capitalist planters and investors who launched each of these antebellum southern industries (transportation, coal, sugar, and textiles) willingly adapted technology to facilitate production. But in all but the case of the steamboats, and despite the acceleration of information exchange, improving transportation systems, and some measure of economic diversification that occurred over the course of the antebellum era, their relative collective unwillingness to view their role in their industry on a larger scale, and well beyond local and regional horizons, prevented greater success and sustained profitability.

Pamela C. Edwards propels us forward into the late-nineteenth- and early-twentieth-century South, where we see a different kind of mentality emerging among southern industrialists. She invites us to look at the significantly different strategies employed by three Carolina textile entrepreneurs to balance their access to abundant, cheap labor and land with their limited capital, technical skills, transportation access, and marketing acumen. She concludes that "the New South did not develop by following any single path to industrialization." Instead, in what marks a sharp turn from antebellum strategies, southern entrepreneurs pursued multiple accommodations with well-established northern financial, business, and technical networks with whom they also had to compete. Though they necessitated a significant measure of compromise, these sets of relationships gave southerners access to critical resources necessary for competition at the national and international level. In each case, entrepreneurs worked with their local community leaders to determine their relative assets and liabilities in their desire to industrialize, and they used that knowledge to form partnerships with experienced northern textile networks eager to pursue interregional and even international development. Moreover, because Edwards finds that each community industrialized according to its own unique set of assets and liabilities, the idea of southern regional economic development remains an egregious misnomer.

Stephen Wallace Taylor tackles this question of regional economic development at mid-twentieth century with his examination of the federally funded Tennessee Valley Authority, which, among other things, delivered cheap electricity to rural residents in one of the most impoverished areas of the nation. Taylor conveys how original TVA board member David E. Lilienthal fought to make the TVA first and foremost a public utility that served the needs of the people as described in his landmark treatise, *TVA: Democracy on the March*. Lilienthal put great faith in the ability of "the experts" to pursue and safeguard the needs of the people. Instead, Taylor argues, the TVA devolved into a "technology-driven bureaucracy" that refused to be answerable either to the local populace or to national environmental critics. By the 1970s, the TVA was far from the great technological and social innovator, "the living laboratory," that President Roosevelt had envisioned. But as Taylor convincingly argues, the TVA throughout its history has remained a strong proponent of the idea that technology has the capacity to improve the quality of life in the Tennessee valley and the nation as a whole.

Because industrialization took place later in the South than in the North, southerners have had to adopt multiple strategies to "catch up" with regions and countries with more mature industrial systems and networks. Those strategies have included forming new relationships with successful industrialists with ample financial and technical capital from elsewhere, as in the case of Pam Edwards's Carrboro and Columbia entrepreneurs, and allowing federal government intrusion, as in the case of Stephen Wallace Taylor's Tennessee Valley Authority. Yoneyuki Sugita asks us to extend our geographical definition of "the modern South" to include such Sunbelt states as California, Arizona, and Nevada. But whether or not we accept his expanded map of the South, his analysis of telemedicine as an especially innovative technology in places that are characterized by large rural populations (as still exists in much of the traditional South) bears consideration. Like several other southern states, Arizona's low taxes and cost of living have compelled many international corporations to relocate. These same attributes, combined with a warm climate, have also attracted droves of retirees, transforming Arizona into one of the fastest-growing states. The need for high-quality health care in this high-tech, knowledge-intensive business climate, coupled with a government-business-industry development strategy, has produced innovative electronic delivery of medicine to rural areas. Although telemedicine, Sugita argues, represents the kind of innovative economic development practices that are beginning to take hold in southern states through this kind of cross-collaboration, it has so far failed to rectify the longstanding disparities in health care for the poor, he is quick to note, especially for the Native American and Hispanic communities.

This volume, of course, will not be the last word on technology, innovation, and industrialization in southern history. Too many questions remain unanswered; too many industries are still unexplored. As editors we urge future scholars on this theme to bring new degrees of sophistication to this subject. How does a gendered analysis help us understand these processes better, for example? We certainly need more researchers to look more carefully for the role of women inventors and innovators in the processes of southern industrialization. But we also need more nuanced interpretations of the gendered nature of innovation and technology in the South, as well as its transformations over time. What did innovative ideas and mechanical and managerial skills mean in respect to manhood in a southern culture that traditionally defined masculinity quite differently?[25] Similarly, how did the adoption and success of certain innovations prompt southerners to begin to define the place of men (and women) with technological knowledge in their society? When can we point to the emergence of a recognizable class of skilled mechanics and engineers, and when did their knowledge give them a significant measure of status and authority? When did these men

25. Although she does not explore such questions in reference to the antebellum American South, Ruth Oldenziel's *Making Technology Masculine: Men, Women and Modern Machines in America, 1870–1945* (Amsterdam: Amsterdam University Press, 1999) rightly conveys how notions of masculinity have been so thoroughly knitted into ideas about technology, innovation, and mechanization in modern American culture.

professionalize and was that process different from the way it progressed elsewhere? Certainly in the late antebellum era, nonnative skilled men were not necessarily warmly received by all. They were as likely to be feared as closet abolitionists as they were valued for their mechanical understanding. Likewise, how can we better include the contributions of laboring people, including African American men and women, to technological change in the South, rather than emphasizing larger processes of change that recognize only those at the top of the social scale?

Other parameters of variation invite closer examination, too. Do we find, for example, a new pattern in respect to rural-urban differences in innovation in the South? Louisiana planters were willing adapters of the latest sugar-processing technologies, and the production of electricity for rural residents in the Tennessee valley and telemedicine for rural Arizonians challenge long-held beliefs in innovation and its application as cosmopolitan phenomena. On another track, what is the relationship between patterns of human migration and the degree of innovation at particular moments and in particular places in the American South, and which drives the other? One of the more obvious examples of this relationship is the availability of tractors after World War II and the great numbers of African Americans moving to the North. Certainly the late-twentieth-century migration of senior citizens to the Sunbelt played a critical role in the pursuit of telemedicine as a valuable delivery system for controlling spiralling health care demands that also proved cost-efficient, even profitable.

Technology has become part of "the narrative" of modernity. It has often been used to symbolize Western superiority. The South, because it has long been romanticized (or criticized) by many different groups as the last region in the country to retain its agrarian roots, is often viewed as a region that was antithetical to technology and innovation, and therefore the "black sheep" in regard to advancing U.S. economic development. This volume offers a more challenging and sophisticated interpretation. Technology and innovation were never absent from southern industrialization. Instead, their benefits had to be carefully weighed in relationship to the costs incurred for the industry and even the larger society. Again and again the priorities of that larger society had a tremendous impact on which technologies and innovations would be incorporated and whether they would be successful. Hence the presumption that southerners saw little or no value in pursuing new technologies and innovations in their industries at any point in the past is an entirely false one. Southerners from the early nineteenth century onward were as eager to embrace technology and innovation as any other Americans; they, too, viewed technology as the path to profit and modernity. What southerners faced in greater quantity than people in the other regions of the United States were significant limitations on what technologies and innovations they were willing to incorporate, especially in a world where slaveholding agriculture before the Civil War shaped the allocation of so many resources. Indeed, the scarcity of capital and the continued reliance on agriculture affected allocation right into the late nineteenth and twentieth centuries. Only with the infusion of federal spending during the depression and World Wars could the industrial South marshal the capital to innovate more fully. The results were significant, with the manufacturing sector contributing mightily to the phenomenonal Sunbelt South boom by mid-century, and the banking and electronics industries into the twenty-first. The South's recent embrace of technology and innovation in the modern era does not mark a radical change from its past. Instead, as this volume demonstrates, longstanding attention to new technologies and the desire to pursue innovation to industrialize were always part and parcel of southern economy and society.

The editors thank all the contributors for their engagement with this volume. We also wish to express our gratitude to Beverly Jarrett, editor in chief at the University of Missouri Press, for her commitment to this book series from its inception, as well as Sara Davis, our fabulous editor. Michele Gillespie also wishes to thank Kate Kemmerer, her former Wake Forest University student assistant, for her invaluable help throughout the genesis of this volume. Finally, we wish to thank Randy Patton, Shaw Industries Chair at Kennesaw State University, for his critical support.

Steamboats and Southern Economic Development

ROBERT H. GUDMESTAD

WHEN THE *NEW ORLEANS* LEFT THE PITTSBURGH WATERFRONT on October 20, 1811, a great crowd cheered her departure. Women waved their handkerchiefs, and men tossed their caps in the air. The boat shoved off into the Monongahela River, went upstream a short distance, and then curved around and disappeared downstream. In many ways the *New Orleans* was typical of craft found on the Mississippi River system. She was made from white pine that had been cut from local forests and floated to the local shipyard. Workers in a huge saw pit had cut the trunks into planks, and local shipwrights had fitted the boards together. The sky-blue boat sported a bowsprit, portholes, and two masts and two cabins. But the *New Orleans* drew a large crowd on that chilly Pennsylvania Sunday because she surpassed her predecessors. She was huge, the largest boat built in Pittsburgh. At 148 feet by 32 feet and 371 tons burden, she was 50 percent larger than most flatboats. What really set the *New Orleans* apart were the paddlewheels mounted on the boat's sides and the steam engine in her hold. The low-pressure engine, the parts for which had been hauled from New York to Pittsburgh in wagons, had two copper boilers and a cylinder thirty-four inches in diameter. The $38,000 price tag also made the *New Orleans* the most expensive boat built in Pittsburgh.[1]

1. J. H. B. Latrobe, *The First Steamboat Voyage on the Western Waters* (Baltimore: J. Murphy, 1871), 9–15; George L. Norton to James A. Henderson, June 6, 1911, and H. Dora Stecker to George M. Lehman, July 7, 1911, both in *New Orleans* (Steamboat) Collection, 1911, Historical Society of Western Pennsylvania, Pittsburgh, Pennsylvania; Jay Feldman, *When the Mississippi Ran Backwards: Empire, Intrigue,*

This first steamboat west of the Appalachian Mountains was the product of northern engineering, manufacturing, and capital. Her investors, however, saw the boat's great potential for the South and intended that she make regular runs between New Orleans and Natchez. The boat immediately proved her worth, prompting a Natchez slave to remark, "By golly, Sa, old Massesseppa got her massa." Southerners immediately realized the value of the steamboat for conquering the swift currents of the western waters and embraced the new technology. They built their own craft and developed extensive shipping networks on the waters that drained into the Gulf of Mexico. Steamboats, although originally associated with Robert Fulton and New York, quickly became symbolic of a southern river culture. Riverboats accelerated the formation of the southern social order and became a catalyst for the development of farms, plantations, cities, industry, and society along the western rivers.[2]

Historians of the American South, though, have been slow to recognize the importance of steamboats. Three newer works on steamers study African American riverboat workers, describe boats in Louisiana, and discuss environmental issues. Most other works on the South, both specialized works and general histories, mention steamboats in passing. The place of steamboats in the transportation revolution of the Early Republic is also a neglected subject. The best work on the matter was done during Harry Truman's presidency, and a later, purely economic, history also exists. An examination of the interaction between steamboats and the antebellum South is long overdue.[3]

Murder, and the New Madrid Earthquakes (New York: Free Press, 2005), 104–31, 148–53; James M. Powles, "*New Orleans:* First Steamboat Down the Mississippi," *American History* (2005): 48–55; Charles W. Dahlinger, "The *New Orleans*," *Pittsburg [sic] Legal Journal* 59 (1911): 579–91; F. Van Loon Ryder, "The 'New Orleans': The First Steamboat on our Western Waters," *Filson Club Historical Quarterly* (1963): 29–37.

2. J. H. B. Latrobe, *A Lost Chapter in the History of the Steamboat* (Baltimore: J. Murphy, 1871), 34 (quotation).

3. Thomas C. Buchanan, *Black Life on the Mississippi: Slaves, Free Blacks, and the Western Steamboat World* (Chapel Hill: University of North Carolina Press, 2004); Carl A. Brasseaux and Keith P. Fontenot, *Steamboats on Louisiana's Bayous: A History and Directory* (Baton Rouge: Louisiana State University Press, 2004); Ari Kelman, *A River and its City: The Nature of Landscape in New Orleans* (Berkeley: University of California Press, 2003), 50-68. Two of the few works that devote more than cursory reference to the relation of steamboats to the antebellum South are John Hebron Moore, *The Emergence of the Cotton Kingdom in the Old Southwest* (Baton Rouge: Louisiana State University Press, 1988), 156–64; and Edith McCall, *Conquering the*

Steamboats were important to the social, demographic, and economic growth of the United States because they brought routine, speed, and reliability to movement. Early-nineteenth-century transportation in the country's interior was fraught with difficulty. Travel by river offered numerous advantages over getting about using roads and canals, but it came with its own troubles. Downstream travel, which constituted 90 percent of Mississippi River traffic in 1810, was accomplished mainly by flatboats. These "wooden prisons" made their appearance on the western waters around the time of the American Revolution and were often assemblages of rough logs hastily nailed together. Their rudders and oars were of little help in swift currents, as their bulkiness left them at the mercy of nature, especially when they took on large amounts of cargo. A federal investigation concluded that flatboats could be managed "only slowly and with difficulty." Although cheap to build, flatboats held relatively small amounts of freight. The first boats carried no more than 40 tons while later versions transported up to 150 tons of freight. Flatboats were disposable watercraft. Once they reached New Orleans, their crews broke them apart, sold the wood, and normally walked back home.[4]

Upstream travel was even worse. It was a slow, painful process that reduced commercial activity to a trickle. Keelboats, each of which could carry from 40 to 100 tons of freight, became common around 1780. Unlike the boxlike flatboats, keelboats resembled sailing ships. They were meant to cut through the water rather than float on it. Sinewy and callused crews overcame the river's powerful current whenever possible through sails (which were unreliable),

---

Rivers: Henry Miller Shreve and the Navigation of America's Inland Waterways (Baton Rouge: Louisiana State University Press, 1984). For steamboats in the antebellum era, see Louis C. Hunter, *Steamboats on the Western Rivers: An Economic and Technological History* (1949; reprint, Mineola, N. Y.: Dover Publications, 1993); Erik F. Haites, James Mak, and Gary M. Walton, *Western River Transportation: The Era of Early Internal Development, 1810–1860* (Baltimore: Johns Hopkins University Press, 1975); and Adam I. Kane, *The Western River Steamboat* (College Station: Texas A & M Press, 2004).

4. Kelman, *A River and its City,* 52 (first quotation); Haites, Mak, and Walton, *Western River Transportation,* 14 (second quotation); Michael Allen, *Western Rivermen, 1763–1861: Ohio and Mississippi Boatmen and the Myth of the Alligator Horse* (Baton Rouge: Louisiana State University Press, 1990), 74–80; W. Wallace Carson, "Transportation and Traffic on the Ohio and the Mississippi before the Steamboats," *Mississippi Valley Historical Review* 7 (June 1920): 26–38; Leland D. Baldwin, "Shipbuilding on the Western Waters, 1793–1817," *Mississippi Valley Historical Review* 20 (June 1933): 29–44.

oars (which were ineffective), or long poles (which had limited usefulness). When these methods failed, crews resorted to cordelling, where a sailor swam to shore with a thick, heavy rope called a cordelle. Other members followed and then pulled the ship upriver by trudging along the shore, dragging the rope with them. Should the bank prove slippery or uneven, crews turned to warping, where they fastened the cordelle to a tree and then hauled it in, thus propelling the boat forward in a tedious tug-of-war against the current. A keelboat might travel thirty miles on a good day, but most crews managed only half that distance before the sun set. The 1,350–mile journey from New Orleans to Louisville typically took three to four months. These primitive conditions limited upstream freight on the Mississippi River to not more than 6,500 tons a year. Steamboat tonnage provides a measure of comparison to show how the expense and difficulty of keelboat travel kept the supply of goods from outside the region at low levels. In 1820 western river steamboats had a total carrying capacity of 13,890 tons. If these boats made only one trip per month, they could carry twenty-five times more freight than went upriver in 1811.[5]

The restrictive effects of poor transportation led southerners to experiment with new forms of transportation. In 1802 Louis Valcourt of Kentucky and James McKeever of Philadelphia devised a plan for a steam-powered river vessel that could travel between New Orleans and Natchez. They purchased a steam engine in Philadelphia, shipped it to New Orleans, and fitted it into a boat. Before the craft became operational, the Mississippi River flooded and pushed it into a remote swamp. Valcourt and McKeever, who had already sunk $15,000 into the project, needed a quick infusion of capital. They borrowed money from William Donaldson, who secured use of the engine to power his sawmill until a new hull was ready. The sawmill was a commercial success until it went up in flames, ending this first attempt to build a steamboat on the western waters.

In 1807 some enterprising New Orleans residents built the "horse boat" for which horses walked on a type of treadmill that turned a

---

5. Allen, *Western Rivermen,* 69–73; Kelman, *A River and its City,* 51; Haites, Mak, and Walton, *Western River Transportation,* 12-26; Herbert Quick and Edward Quick, *Mississippi Steamboatin': A History of Steamboating on the Mississippi and Its Tributaries* (New York: Henry Holt, 1926), 12–19. The statistics are drawn from James Hall, *The West: Its Commerce and Navigation* (1848; reprint, New York: Burt Franklin, 1970), 130; Hunter, *Steamboats on the Western Rivers,* 33.

set of paddle wheels. The boat left for Louisville but never made it. Near Natchez, according to a New Orleans paper, "some twelve to twenty horses were used up on the treadwheel" and the trip was abandoned.[6]

The planning behind the *New Orleans* proved superior to these failed attempts. Robert Fulton, who had built the first commercially viable steamboat, was one of the boat's investors. His ambitions were too great to be confined to New York. He wrote his business partner Robert Livingston, a signer of the Declaration of Independence and negotiator of the Louisiana Purchase, that "Everything is completely proved for the Mississippi, and the object is immense." Livingston agreed, and recommended they hire Nicholas Roosevelt to scout the Mississippi River and supervise construction of their new steamboat. Roosevelt, Theodore Roosevelt's granduncle, went to Pittsburgh in 1809 and procured a flatboat. He then took his young wife on a trip downriver to New Orleans. When Roosevelt returned to New York and endorsed the possibility of a steamboat for the western waters, Fulton and Livingston went public with their plans. Their venture, the Ohio Steam-Boat Navigation Company, announced that it had been "exploring and ascertaining the practicability of navigating the Ohio, and Mississippi rivers, by actual experience" and opened its stock subscription books in Pittsburgh, Cincinnati, and Louisville. The company's incorporators were some of the most powerful men in New York: Fulton, Livingston, Roosevelt, Daniel D. Tompkins (governor of New York and later vice president under James Monroe), and DeWitt Clinton (lieutenant governor of New York and later the primary enthusiast for the Erie Canal).[7]

Even before Roosevelt arrived in Pittsburgh to supervise construction of the *New Orleans*, Fulton and Livingston met with Governor William Claiborne of Louisiana. They relayed their plans

6. S[arepta] Kussart, *Navigation on the Monongahela River* (n.p., 1971), 363; John Hebron Moore, *Andrew Brown and Cypress Lumbering in the Old Southwest* (Baton Rouge: Louisiana State University Press, 1967), 13–4; McCall, *Conquering the Rivers*, 88; *New Orleans Gazette*, July 23, 1807, as quoted in Mak, Haites, and Walton, *Western River Transportation*, 17 (quotation); Gould, *Fifty Years*, 40.

7. Fulton as quoted in Kirkpatrick Sale, *The Fire of His Genius: Robert Fulton and the American Dream* (New York: Free Press, 2001), 14; *Baltimore Whig*, October 20, 1811, as reprinted in *Cincinnati Liberty Hall*, November 20, 1811, as quoted in Ethel C. Leahy, *Who's Who on the Ohio River and its Tributaries* (Cincinnati: E. C. Leahy, 1931), 302. The text prints Tompkins's name as David D. Tompkins, a mistake made in other contexts.

to Claiborne and promised that steamboats would reduce shipping costs and traveling time by one quarter. Fulton and Livingston also explained how expensive it was to build the new boats and promised to maintain consistent river traffic in exchange for the exclusive right to navigate steamboats within the limits of Louisiana. Claiborne, a native southerner, believed in the potential of steamboats to transform New Orleans into the greatest city west of the Appalachians. He prevailed upon the legislature to grant a monopoly for the company to operate steamboats within the Territory of Orleans, soon to become the state of Louisiana. The monopoly would commence January 1, 1812, and last eighteen years.[8]

Although the monopoly was important, the success—and profits—of the Ohio Steam-Boat Navigation Company hinged on a successful demonstration of steam power on the Ohio and Mississippi rivers. The voyage of the *New Orleans* was part experimentation and part advertisement for the future services of the boat. Roosevelt understood this fact when he and fourteen others left Pittsburgh. One of the passengers was Mrs. Roosevelt, who was great with child and added another member to the expedition in Louisville. Traveling at a robust eight miles per hour, the boat steamed to Cincinnati in two days and to Louisville in two more. She also became the stuff of legend. According to an account written fifty years after the fact, when a Kentucky farmer heard the puffing of the *New Orleans*, he warned his neighbors that the British were invading. They rushed to the riverbank in time to see "the monster of the waters" pass. Another dubious story tells how when the boat arrived in Louisville at night, she released steam from her boilers. The noise frightened some of the town's residents into running into the woods lest a comet fall from the sky and obliterate them. Accurate or not, these stories convey the sense of novelty and wonder isolated farmers of river communities felt upon seeing the *New Orleans*.[9]

Roosevelt had to wait at Louisville since water was only trickling over the Falls of the Ohio. He took the opportunity to steer the boat back to Cincinnati and amazed the locals by going upstream without poles, paddles, or sails. In late November the Ohio River rose

8. "From Robert Fulton," *Waterways Journal*, January 28, 1899, 3; Dahlinger, "The *New Orleans*," 588; Caroline S. Pfaff, "Henry Miller Shreve: A Biography," *Louisiana Historical Quarterly* 10 (April, 1927): 215-6.

9. "First Western Steamboat," *Waterways Journal*, November 7, 1896, 8; Powles, "*New Orleans*," 51-2; Feldman, *When the Mississippi Ran Backwards*, 126 (quotation).

enough for the *New Orleans* to scrape over the falls and continue her journey. A bit further downstream, the boat moored to an island, only to have the land disappear during the New Madrid earthquake. A wall of brown water nearly swamped the boat, and the violent shaking washed trees and huge chunks of land into the river. All through the night, aftershocks roiled the waters, tossing the boat and sickening the passengers. A nasty sulfur smell hung in the air. The morning revealed an eerie scene: huge cracks in the earth, trees choking the river, and traditional landmarks obliterated. The *New Orleans*'s pilot and crew had to cope with this new and confusing reality, but they arrived in New Orleans on January 10, 1812.[10]

The *New Orleans* created an immediate sensation in Louisiana. John L. Lewis, remembering the boat's arrival many years later, said the legislature adjourned to see the boat arrive at the levee. One resident noted how Roosevelt "has been astonishing and amusing the multitude" by charging passengers three dollars to ride ten miles downriver to English Turn and back. The boat, this businessman confided, exceeded the expectations of even the most positive booster. Another New Orleans resident gushed that the riverboat "is so well addapted to the trade of the Mississippi" that its prospects were limitless. Roosevelt hoped to convert some of this early enthusiasm into cash when he opened the stock subscription books for new investors. The Ohio Steam-Boat Navigation Company had grand plans and needed a steady infusion of cash to realize its dreams. The company wanted to set up a system of relays between cities (New Orleans and Natchez, Natchez and Louisville, Louisville and Pittsburgh) in an early version of the packet system. It was building two more boats, the *Vesuvius* and the *Aetna*, and had plans for more.[11]

10. Latrobe, *First Steamboat Voyage*, 21–5; Ryder, "First Steamboat," 33–36; "On the Old Missouri," *Waterways Journal*, December 13, 1902, 3; "Commencement of Navigation," *Waterways Journal*, January 4, 1896, 3; Powles, "*New Orleans*," 51-2; Feldman, *When the Mississippi Ran Backwards*, 148–50.

11. Emerson W. Gould, *Fifty Years on the Mississippi; or, Gould's History of River Navigation* (St. Louis: Nixon-Jones, 1889), 240; William Kenner to Stephen Minor, January 20, 1812, Minor Family Papers, Southern Historical Collection, University of North Carolina, Chapel Hill, North Carolina (first quotation); Reuben Kemper to Edward Livingston, January 21, 1812, as quoted in Adam Rothman, *Slave Country: American Expansion and the Origins of the Deep South* (Cambridge, Mass.: Harvard University Press, 2005), 81 (second quotation); *New Orleans Monitor*, March 5, 1812, as quoted in Dahlinger, "The *New Orleans*," 587; Pfaff, "Henry Miller Shreve," 216.

The company was off to a good start. Published accounts reveal that the *New Orleans* carried enough passengers and freight to make a $20,000 profit its first year, a "revenue superior to any other establishment in the United States." Most significant, perhaps, it enjoyed the goodwill of the many residents of the Mississippi River valley. The *Louisiana Courier* observed that the "privileges" granted Fulton and Livingston "contain nothing which may be considered 'odious.'" Monopolies tend to offend modern sensibilities, but they were not so controversial in the early nineteenth century. The North Carolina legislature extended a monopoly to John Stevens in 1812, and Fulton, of course, had monopoly rights in New York. The government could not, or would not, fund technological innovation, but it could create conditions where it might thrive. Entrepreneurs wanted the incentive of profit, and that often meant a limited range of protection from competition. As the *Courier* put it, "the results of their labors should go 'to the benefit of the investors' as well as the communities enjoying them." For this particular editor, at least, there was no conflict between a monopoly and the public good. The territorial legislature, moreover, did not grant carte blanche to the Ohio Steam-Boat Navigation Company. Its freight charge was established at no more than three-quarters the average rate charged by other mercantile boats on the interior waters.[12]

The company's monopoly was soon tested as enterprising westerners hastened to build their own boats. Christian Wilt, a St. Louis merchant, mulled over whether to buy a steamboat in 1813. The asking price, he discovered, for a boat with a forty-ton displacement was about $7,000. That figure seems ridiculously low in comparison to the cost of the *New Orleans*, and perhaps was one reason Wilt thought the boat would bring him "a great profit indeed." He rejected the scheme because of problems securing funds, but Daniel French and Samuel Smith built their own boat. French and Smith, both from Pennsylvania, constructed the *Comet*, a small sternwheel craft. The boat's capacity of twenty-five tons was too small to make

12. Zadok Cramer, *The Navigator: Containing Directions for Navigating the Monongahela, Allegheny, Ohio, and Mississippi Rivers*, 11th ed. (Pittsburgh: Cramer, Spear & Eichbaum, 1821), 13 (first quotation); *Louisiana Courier*, March 11, 1812, as quoted in Kussart, *Navigation on the Monongahela River*, 368 (remaining quotations); Dahlinger, "The *New Orleans*," 588; Alan D. Watson, "Sailing under Steam: The Advent of Steam Navigation in North Carolina to the Civil War," *North Carolina Historical Review* 75 (January 1998): 30–1. For the opinion that the Louisiana monopoly was controversial from the beginning, see Kelman, *A River and Its City*, 50–4.

much of a profit, but her high-pressure engine was an improvement over the Ohio Steam-Boat Navigation Company's boats. The *Comet* made a few trips between New Orleans and Natchez before a planter purchased her and put the engine in a cotton gin. Her limited success persuaded Henry M. Shreve of Pennsylvania to join with Smith to create a bigger and better boat, the *Enterprise.* This boat was also intended to confront directly the Ohio Steam-Boat Navigation Company's monopoly and compete for the right to trade in New Orleans.[13]

Shreve steamed the *Enterprise* from Pittsburgh to New Orleans in 1814. After several unusual events, including the boat carrying ammunition for General Andrew Jackson and the arrest of Shreve, a lawsuit involving the boat triggered a dismissal of the Ohio Steam-Boat Navigation Company's monopoly. The result was a wide-open race for mastery of the western waters and a scramble to get boats into the water. Contemporary accounts reveal that even before the monopoly was officially struck down, boats were built all along the southern waterways. In 1815, for instance, David Prentice built the *Pike* at Hendersonville, Kentucky, while John K. West and Francis Duplisses built the *Louisiana* in New Orleans three years later. Between 1811 and 1821, southerners in thirteen different towns manufactured at least twenty-eight steamboats. Most of these efforts were clustered in Kentucky, and one jerry-rigged creation supposedly sported an engine from a flour mill and belched cinders and ashes from its brick smokestack. Not all of the early boats were so hastily engineered, of course, but the story is suggestive of the builders' desire to get boats into the water as quickly as possible.[14]

13. Marietta Jennings, *A Pioneer Merchant of St. Louis, 1810–1820* (New York: Columbia University Press, 1939), 132 (quotation); Hunter, *Steamboats on the Western Rivers,* 13–15; "First Steamers on the Mississippi," *Waterways Journal,* August 31, 1895, 12; "Fulton and the Steamboat," *Waterways Journal,* February 4, 1899, 3; "Life of Capt. C. W. Batchelor," *Waterways Journal,* August 8, 1896, 6; "Commencement of Navigation," *Waterways Journal,* January 4, 1896, 3; McCall, *Conquering the Rivers,* 96–9; Albert A. Fossier, *New Orleans: The Glamour Period, 1800–1840* (New Orleans: Pelican Publishing Company, 1957), 30.

14. Unknown date in July, 1818 *Louisiana Gazette,* as quoted in "First Steamers on the Mississippi," *Waterways Journal,* August 31, 1895, 12; "Life of Capt. C. W. Batchelor," *Waterways Journal,* August 8, 1896, 5; Hall, *The West,* 128; Fossier, *New Orleans,* 30; George H. Yater, *Two Hundred Years at the Falls of the Ohio: A History of Louisville and Jefferson County* (Louisville: Heritage Corporation, 1979), 34–35; Capt. F. L. Wooldridge, "Iron Men - Wooden Boats," Scrapbook volume 1, Jones-Wright Steamboat Collection, Special Collections, Howard-Tilton Memorial Library, Tulane

These early efforts to build steamboats display a high degree of cooperation between the North and the South. Many of the steam engines came from New York, at first the only state with foundries large enough to forge such machinery. But the other elements required to make a steamboat were not exclusive to the North. Capital, expertise, and ingenuity were available in the West and South. Edward Livingston even proposed moving "our Workshop to some place below the falls" at Louisville. The abundance of good wood, he wrote to Fulton, made the mouth of the Cumberland an advantageous place to situate their western operations. Once steamboats demonstrated they could handle the strong currents of the Mississippi River, most southerners embraced the new technology, and others directed their investments into the new technology. Residents along the major rivers of the United States knew the importance of steamboats in a deeply personal way, and inventive southerners who relied on the utility of riverboats were simply part of a larger American trend toward industrialization and technological innovation. As we shall see, southerners were able to direct the use of steamboats in ways that usually bolstered their slave society and plantation culture. Southern ingenuity, ultimately, had to preserve rather than challenge slavery.[15]

Southerners who could not arrange to build boats devised ways to purchase them. Businessmen in Natchez, Nashville, and Louisville formed companies that put boats on the murky waters of the Mississippi. Records survive for the Natchez investors, who formed the Natchez Steam Boat Company on July 4, 1818. It was a joint stock venture to be capitalized at $100,000. The investors were the well-heeled elite of Natchez—planters, presidents of banks, magistrates, attorneys, merchants, and the man who would become

---

University, New Orleans, Louisiana, hereinafter Tulane; Hunter, *Steamboats on the Western Rivers,* 12-20; Moore, *Emergence of the Cotton Kingdom,* 157–8; McCall, *Conquering the Rivers,* 109–35; Kussart, *Navigation on the Monongahela River,* 377. The southern towns manufacturing steamboats were Virginia: Wheeling; Kentucky: Brandenburg, Frankfort, Leestown, Louisville, Maysville, Newport, Portland, Salt River, Shippingport; Alabama: Blakely, Ft. Stephens, Mobile; Louisiana: New Orleans.

15. Edward Livingston to Robert Fulton, January 14, 1814, Steamboat Collection, Missouri Historical Society, St. Louis, Missouri. For more discussion of technology and its relation to accepted American values, see Leo Marx, *The Machine in the Garden: Technology and the Pastoral Ideal in America* (New York: Oxford University Press, 1964) and John F. Kasson, *Civilizing the Machine: Technology and Republican Values in America, 1776–1900* (New York: Hill and Wang, 1976).

Mississippi's first state governor. When the company purchased Fulton and Livingston's first two boats, it was symbolic of how the old aristocracy of the East had given way to the brash businessmen and planters of the South. The profiles of investors in other southern cities highlight the involvement of successful merchants, planters, or attorneys, many of whom later became part of the ruling elite. For instance, a future governor of Tennessee, William Carroll, helped incorporate the Nashville Steamboat Company.[16]

These early steamboat entrepreneurs were eager to exploit a perceived advantage in technology to earn a profit. They recognized the ability of steamboats to link them to regional, national, and international commercial networks, and they did not fear connections to more markets. The degree to which the South's economy was capitalist is a matter of some debate, but southerners tended to invest in land and slaves rather than in banking and industry. Staple-crop production yielded proven dividends and tended to blind southerners to other economic possibilities. Steamboats, though, were a good fit for southern businessmen, who wanted their capital invested in something that would enhance the value of land and slaves. The relatively low cost of a riverboat, perhaps eighty dollars per measured ton, could be financed by either a few individuals or through stock sales. Other expenses like supplies, maintenance, and wages were low and were paid out over time. No supporting infrastructure was necessary since the rivers and city levees were already in place. Steamboats also turned a quick profit. Once completed, they generated immediate revenue and could be free of debt in a few years. Steamboats supplemented the agricultural economy by becoming the primary carrier of staple crops west of the Appalachian Mountains and easily meshed with the plantation system.[17]

16. Natchez Steam Boat Company agreement, Natchez Trace Steamboat Collection, Center for American History, The University of Texas at Austin, Austin, Texas, hereinafter UTX; "Iron Men -Wooden Boats" vol. 1, Tulane; Yater, *Two Hundred Years*, 34–35; Paul H. Bergeron, Stephen V. Ash, and Jeanette Keith, *Tennesseans and their History* (Knoxville: University of Tennessee Press, 1999), 113; Anita Shafer Goodstein, *Nashville 1780–1860: From Frontier to City* (Gainesville: University of Florida Press, 1989), 35. The occupations of the individuals in Natchez were found by comparing the names on the agreement with information in D. Clayton James, *Antebellum Natchez* (Baton Rouge: Louisiana State University Press, 1968).

17. Hunter, *Steamboats on the Western Rivers*, 110–1; George Rogers Taylor, *The Transportation Revolution, 1815–1860* (New York: Rinehart, 1951), 69–70.

This integration into more sophisticated commercial networks was not without its dangers, and southerners typically recognized potential perils that accompanied investment in steamboats. Fires, collisions, snags, or explosions could bring sudden ruin to investors. John Summerville of the Nashville Steamboat Company began to "dread" steamboat ownership. At one point he nervously reminded a business associate that two riverboats had sunk recently, and that "scarcely one makes the voyage without being perforated by the snags in the rivers." Still, Summerville thought the rewards—the "profits are great indeed," he observed—worth the risks. He put his finger on the essential point: the money to be made in the steamboat business was a powerful incentive and enough to attract investors across the South.[18]

Downstream commerce experienced substantial changes from the advent of steamboat service. Steam power did not put flatboats and keelboats out of business, but it did effect a startling shift in what they carried and where they operated. Some flatboats carried all types of material, but most shifted to transporting pork and tended to stay in the upper Ohio Valley. In their place, steamboats prowled the lower Mississippi, its tributaries, and the rivers that wound through Alabama and Mississippi. Some boats made the long run between New Orleans and Nashville, but even more found their niche in the packet trade, or the shuttling between two close cities. Intense competition forced a moderate reduction in downstream shipping costs. The charge for sending freight downstream in 1810 was about a dollar per one hundred pounds. That figure dropped to nearly sixty cents in the mid 1820s and to fifty cents a decade later. While not as startling as the reductions in upstream costs, the savings were still useful. (See Table One for detailed information.)[19]

Steamboats most obviously changed the nature of upstream commerce and travel. Before riverboats, upriver freight rates were

18. John Summerville to Thomas E. Sumner, June 30, 1818, John Sumner Russwurm Papers, Tennessee State Library and Archives, Nashville, Tennessee, hereinafter TSLA, (quotations). For a discussion of southern planters as entrepreneurs, see James Oakes, *Slavery and Freedom: An Interpretation of the Old South* (New York: W. W. Norton, 1990), 40–79. I have found no contemporary apprehension or criticism regarding the changing commercial relations in the South. On the contrary, all evidence is of southerners who welcomed the connection to distant markets.

19. Haites, Mak, and Walton, *Western River Transportation*, 150–57; Allen, *Western Rivermen*, 140–7.

Table 1.
Average Steamboat Freight Rates in the Louisville to New Orleans Trade

PERIOD	UPSTREAM	DOWNSTREAM
Before 1820	$5.00	$1.00
1820–29	1.00	0.625
1830–39	0.50	0.50
1840–49	0.25	0.30
1850–59	0.25	0.325

Adapted from Haites, Mak, and Walton, *Western River Transportation*, 157.

about five dollars per one hundred pounds of freight. That number had shrunk to a dollar by the mid 1820s and to about fifty cents a decade later. By 1840 it cost about the same to ship material up- or downstream (Table One). The whole Mississippi valley suddenly had access to a wider array of affordable necessities and luxuries. Southerners took advantage of this revolution in commerce and used steamboats to supply their farms and plantations with a dizzying array of items. In 1824 J. T. McNeil, for instance, bought a cotton gin from William Kenner and Company in New Orleans and shipped it on the *Louisiana* for ten dollars. Items shipped on steamboats ranged from slaves, rope, kegs of nails, potatoes, mules, hay, and codfish to candy, oysters, figs, and pianos. Steamboats were much more convenient and cost-effective than keelboats. They brought supplies directly to plantations by stopping at the landings or unloaded their cargoes in the river towns. In time, southerners became dependent on steamboats for almost everything created outside the region, and they were closely tied to the wider commercial world of the Atlantic Ocean. Within a decade of its introduction, "the steamboat ceased to be a novelty on the Mississippi," according to the *New Orleans Gazette*, "and because of this, became a recognized agent of commerce of the valley."[20]

20. Haites, Mak, and Walton, *Western River Transportation*, 150–57; *Louisianias* Records, Tulane; *New Orleans Gazette*, June 12, 1818 as quoted in Fossier, *New Orleans*, 31. The listing of the items shipped on steamboats is taken from an examination of the receipts in the Natchez Trace Steamboat Collection, UTX. Steamboats were not as influential on the rivers of the southern states along the Atlantic seaboard. The rivers were not as deep nor as long as the western waters. For instance, regular steamboat service in North Carolina was limited to a one hundred mile stretch of the Cape Fear River between Wilmington and Fayetteville. See Watson, "Sailing under Steam," 52.

By the 1830s, steamboats had become the workhorse of the plantation regime thanks to improved designs and advances in technology. They enabled western planters to ship cash crops like cotton, sugar, and tobacco much more efficiently, safely, and economically. It is no coincidence that a cotton bale was the first piece of freight carried on a Mississippi steamboat. William Kenner saw the *New Orleans* arrive in 1812 and immediately realized that he should follow suit and begin shipping cotton on steamboats. Kenner was a shrewd New Orleans factor, commission merchant, and planter who handled the business affairs of planters up and down the Mississippi River. One of his clients was also his father-in-law: Stephen Minor, a wealthy Natchez planter. Kenner wrote Minor that he had spoken "to Mr. Roosevelt on the subject of taking down your cotton next trip." Kenner added that he was sending a shipment of oysters to Natchez, certainly a highly unusual delicacy away from the Gulf Coast and one that would probably spoil on a flatboat. It is a measure of Kenner's dissatisfaction with the flatboat trade that he was willing to rely upon an unproven technology for the movement of Minor's most valuable product.[21]

Yet another advantage of steamboats over flatboats and keelboats in shipping cash crops was their size. A typical flatboat could carry no more than 150 bales of cotton, a puny number when compared to later steamboats. The production of sugar, tobacco, or hemp was similarly hampered by limited transportation opportunities. Keelboats were no larger and were more problematic; they were long and narrow, making it difficult to stack bales, barrels, or crates very high. For a planter to expand his cultivation he also had to increase his construction of flatboats or arrange for more keelboats to stop at his plantation. Steamboats helped raise productivity on plantations by freeing up bond servants to work a staple crop rather than transporting it or building watercraft. Planters always seemed to be labor-hungry during picking season, and steamboats were a welcome way to manage labor more efficiently. As can be surmised from Table Two, steamboats were not solely responsible for the dramatic increase in cotton and staple-crop production. A

21. William Kenner to Stephen Minor, January 10, 1812, as quoted in Leonard V. Huber, "Beginnings of Steamboat Mail on Lower Mississippi," *American Philatelist* 74 (December 1960): 189. For a contemporary of Kenner who was also dissatisfied with flatboat service, see Lewis E. Atherton, "John McDonogh and the Mississippi River Trade," *Louisiana Historical Quarterly* (January 1943): 37–43.

Table 2.
Estimates of steamboat tonnage on Western Rivers and freight shipped through New Orleans

YEAR	STEAMBOAT TONNAGE ON WESTERN RIVERS	FREIGHT SHIPPED THROUGH NEW ORLEANS (TONS)
1820	14,208	106,700
1825	12,527	176,400
1830	24,574	260,900
1836	57,367	437,100
1840	82,626	537,400
1845	96,155	868,000
1850	134,566	886,000
1855	172,695	1,247,200
1860	195,022	2,187,600

Adapted from Haites, Mak, and Walton, *Western River Transportation*, 124, 132.

steady white migration, the interstate slave trade, and Indian removal were also important factors in the development of the Old Southwest. Steamboats, however, became a major component in the growth of plantations.[22]

Steamboats seemed to be designed specifically for the transportation of cotton. The peculiar geography of the West was the greatest influence on the boats' structure, of course, but it is striking how they so completely accommodated the cotton traffic. Their hulls were commonly no more than ten feet below the surface, allowing the boats to operate in the shallow drafts of western waters. Designers widened and lengthened the decks to create stability and provide more cargo space but still enable the vessels to operate in shallow waters. On each boat, the guards, those portions of the main deck that extended beyond the hull, provided a measure of protection for the paddle wheels and facilitated the movement of passengers and freight on and off the boat. Next came the staterooms, texas deck, and pilot house stacked on top of one another, like layers of a wedding cake. With the pilot and captain at the apex, a steamboat could pile an immense amount of cotton on the main deck and not impede navigation or maneuver. James A. Watkins

22. Allen, *Western Rivermen*, 74–76; James S. Buckingham, *The Slave States of America*, 2 vol. (London, 1842), 1:268; Walter Prichard, Fred B. Kniffen, and Clair A. Brown, eds., "Southern Louisiana and Southern Alabama in 1819: The Journal of James Leander Cathcart," *Louisiana Historical Quarterly* 28 (July 1945): 824.

recalled an 1826 trip on which about 400 bales of cotton bound for New Orleans were stacked on the *Fort Adams.* The produce was so heavy that the crew placed logs under the guards to steady the craft. Stacking the 400 to 500–pound bales as high as physically possible became common practice on western boats, and photographs from the Civil War era show ships groaning under the weight. In 1836, for instance, the *Daniel Webster* carried 2,375 bales of cotton out of Nashville. Carrying huge amounts of cotton became a badge of honor, as ambitious captains bragged that their steamboats were "loaded to the guards." Eventually the cotton trade became so lucrative that some steamboats specialized in hauling the staple. The *Ozark* advertised itself as a "cotton boat" that would "pay strict attention to the calls of the planters." The amount of cotton shipped through New Orleans bears silent witness to steamboats' influence. In 1816, a time when steam power was just becoming known, 37,000 bales landed in the Crescent City. Six years later, the figure was 161,000, and it was 428,000 in 1830.[23]

Cotton cultivation was not the only agricultural pursuit that benefited from contact with steamboats. Sugar, hemp, and tobacco plantations might have reaped even more benefits. Whereas cotton producers relied on riverboats to send cotton in one direction—to New Orleans for reshipment to northern mills or English markets— the producers of sugar and tobacco sent their products up and down the river. Steamboats enabled planters to plug into regional markets.

The sugar masters of Louisiana, for instance, could ship sugar or molasses to Omaha, St. Paul, Louisville, Pittsburgh, or virtually anywhere in between. James Dalzell, who cultivated sugar in Louisiana, shipped 300 barrels of molasses on the *Saxon* to Pittsburgh. Most of the Pittsburgh-made barrels were reshipped at Louisville, but a Pittsburgh grocer also explained that some cargo traveled in one boat from the New Orleans region to Pittsburgh. Planters exploited the advantages of unimpeded commerce on the western rivers to change the direction of internal trade. Instead of having to ship their goods through New Orleans and New York and

23. James A. Watkins Manuscripts, Tulane; Logbook of Natchez Landings, 1835–1837, entry of May 2, 1839 (first quotation) and November 24, 1836, UTX; *Arkansas State Gazette,* May 30, 1838 (remaining quotations); Stuart Bruchey, ed., *Cotton and the Growth of the American Economy: 1790–1860* (New York: Harcourt, Brace and World, 1967), 80 (statistics).

then across the mountains, they sent their products upstream. Merchants and commission agents in the Upper South became liaisons between planters and northern buyers. Courtenay and Company of Louisville are just one example. They arranged for the shipment of 143 hogsheads of sugar and 90 barrels of molasses for a sugar planter. Steamboats provided opportunities for southerners to make more nuanced economic decisions by allowing canny southern planters and businessmen to track prices and calculate the best advantage for their products.[24]

The close association of steamboats with the plantation economy is suggested by an undated agreement among a number of residents of Baton Rouge. Twenty-four men, all specifically identifying themselves as planters, established a steamboat company. In a type of prospectus, the men complained of the "extravagant" freight charges and irregular service of steamboats. They hoped to raise twenty-five thousand dollars to build and operate a riverboat that would be more responsive to their needs. In essence, they wanted to bring control to their plantation system by creating an early form of vertical integration. Certainly these early investors expected their vessels to earn profits by carrying passengers and goods of all types up and down the river. Planters specifically, though, came to depend on steamboats in ways unthinkable in 1810. They recognized the value of steamboats in expanding their cultivation activities and marketing their crops. Anthony Gale, a steamboat captain who formed his own boat company, knew how much planters wanted the new technology. In 1816 Gale informed a friend that he had purchased a steamboat and would soon be making regular runs to and from New Orleans. Gale mused that his boat would "be happy news to all the planters." Within a decade after their introduction, steamboats had become indispensable to plantations by hauling staple crops, transporting supplies, and serving as conduits of information.[25]

24. *James Dalzell v. Steamboat Saxon*, #3269, 10 La. Ann (n.s.) 280, (1855) and *Courtenay et al vs. Mississippi Marine and Fire Insurance Company*, #3152, 12 La. Ann. (o.s.) 233, (1838), both in Earl K. Long Library, University of New Orleans, New Orleans, Louisiana, hereinafter UNO; James S. Buckingham, *The Slave States of America*, 2 vol. (London: Fisher, 1842) 1:398–9; Philip Paxton [S. A. Hammett], *Stray Yankee in Texas* (New York: Redfield, 1859), 403; William Cage to James Winchester, April 7, 1822, James Winchester Papers, TSLA.

25. Undated agreement of the residents of Baton Rouge and Bonnete Squarre (first quotation), and Receipt of the *Volcano*, July 26, 1819, both in Natchez Trace

Planters also used steamboats to maximize their control of slave labor. Slave owners often rented out slaves as a means to generate extra income, and it is possible that more slaves were rented in any given year than were sold. Steamboats became one way to move these rented bond servants to their destinations and greatly increased the potential hiring markets. No longer were slaves leased out within walking distance of their homes; they could be sent to distant locales. Henry Summers owned three slaves he normally hired out to a tobacco factory in New Orleans. When work at the factory ceased, he sent the men via the steamboat *Emperor* to cut sugarcane. Likewise, Rachel O'Connor, who lived near St. Francisville, Louisiana, sent "five young negro men" by steamboat to a sugar plantation. Riverboats helped make slave labor more portable and flexible, thereby increasing both the value of slaves and the profitability of plantations.[26]

Steamers also changed the nature of slave labor by turning some bond servants into lumberjacks. The boats consumed an enormous amount of wood, up to thirty cords a day. They stopped often to take on the fuel and paid anywhere from one to five dollars per cord, with two to three dollars being the norm. Wood sales to steamboats were a tremendous windfall for southern planters along the river. Planters were clearing land for farming as a matter of course, but they now dispatched gangs of slaves to prepare the timber for sale rather than burn it. Felling trees and chopping them into pieces also provided additional labor for slaves in times of slack activity in the fields. Isaac Franklin, the erstwhile slave trader turned planter, reaped immense profits from wood sales. His wood yard, located at the confluence of the Mississippi and Red rivers, yielded ten to twenty thousand dollars annually from 1846 to 1850. Perhaps not all the proceeds came from steamboats, but the yard's location suggests that the majority did. Not all wood sellers in the South were wealthy planters, and small wood yards were sprinkled along the rivers. Small-scale entrepreneurs also benefited from the

---

Steamboat Collection, UTX; Anthony R. Gale to Weeks, July 10, 1816, Hill Memorial Library, Louisiana State University, Baton Rouge, Louisiana (second quotation), hereinafter LSU.

26. Jonathan D. Martin, *Divided Mastery: Slave Hiring in the American South* (Cambridge, Mass.: Harvard University Press, 2004); *H. M. Summers v. United States Insurance, Annuity and Trust Co.*, #5423, 13 La. Ann. (n.s.) 504 (1858), UNO; Rachel O'Connor to David Weeks, November 3, 1833, Weeks and Family Papers, LSU (quotation).

wood trade. Like some of his customers, William Johnson, the noted barber of Natchez, sent slaves to chop wood and sell it to steamboats. Although small slaveholders like Johnson benefited from wood sales, planters typically had more property, better riverside land, and more slaves who could cut timber. The consequence was a further consolidation of the economic strength of the large planters and an additional labor burden for bond servants. Steamboats reshaped the contours of unfree labor.[27]

Steamboats influenced African American life and labor in other ways as thousands of slaves and free blacks toiled on the decks of riverboats. Most enslaved Americans probably worked because their masters forced them to, but a significant number sought out employment on the river so they could get out from under the watchful gaze of their masters, earn pay for themselves, or escape. They shared the motivations of free blacks to learn about the wider world, assert personal autonomy, and track down lost family members. African Americans labored as firemen (workers who tossed wood into huge furnaces and stoked fires), roustabouts (strong men who hauled freight on and off the ship), stewards, barbers, cooks, waiters, and chambermaids. Downriver from Louisville, in fact, many steamboats employed more blacks than whites. In the decade before the Civil War, perhaps as many as twenty thousand African Americans worked on western river steamers.[28]

The mobility and savvy of slaves and free blacks on steamers was alarming to some whites. One master advised a friend to sell a "steamboat slave" who was not accustomed to discipline and would be difficult to control. Such individual acts of agency improved the conditions of countless slaves. The enslaved man who negotiated terms of hire on a riverboat and remained out of sight of his master might be assumed to have challenged slavery. Such was generally not the case, as individual acts of resistance were actually ways to accommodate slaves to conditions of bondage. While the slave gained a measure of autonomy, the master gained a steady income

27. David E. Schob, "Woodhawks and Cordwood: Steamboat Fuel on the Ohio and Mississippi Rivers," *Journal of Forest History* (July 1977): 124–32; William Ransom Hogan and Edwin Adams Davis, eds., *William Johnson's Natchez: The Ante-Bellum Diary of a Free Negro* (Baton Rouge: Louisiana State University Press, 1931), 36–8; Hall, *The West*, 129–30; Wendell Holmes Stephenson, *Isaac Franklin: Slave Trader and Planter of the Old South* (Baton Rouge: Louisiana State University Press, 1938), 106, 114.

28. Buchanan, *Black Life on the Mississippi*, 11, 19–52.

from hired slaves, was not responsible for their food or clothing, and had fairly contented bond servants. James Rudd, a Louisville businessman, seems to have been satisfied to hire out his slaves John, George, and Little Charley on various riverboats. Rudd collected from twenty to forty dollars per month from the slaves, who worked on at least six different boats.[29]

Cities also grew apace of plantations thanks to steamboats. Riverboats literally created thousands of urban jobs in the South. Most western boats were packets, so freight might be loaded and unloaded several times before reaching its destination. This continual reshuffling meant that port towns demanded a steady supply of roustabouts, draymen, dockworkers, commission agents, and the like. Armies of dockworkers swarmed aboard ships and hauled the freight to other boats or dockside warehouses. Louisville was especially favored, since the canal around the Falls of the Ohio was small, and passage over the falls was dangerous in all but the highest water. The city's chaotic waterfront was "lined with hacks, omnibuses, wagons, drays, and all kinds of vehicles, hauling passengers and freight to and from the boats arriving and departing." With so many people choking the levee, other businesses moved in to serve them. "Stores, barrooms, lodging-houses, and groceries lined the river front," according to a newspaper story, "and, what with the steamboatmen, flat-boatmen, passengers, teamsters and stragglers, the entire wharf front was a daily scene of business life and activity." The same could be said for the waterfronts of most other southern towns along the river. In Natchez, for instance, hotels that once stood along the Natchez Trace moved to the waterfront in an effort to capture more business.[30]

The construction of steamboats also stimulated industrial production in the South. Many of the early boats were built across the region and created jobs in river cities. But the southern steamboat

29. R. C. Ballard, as quoted in Buchanan, *Black Life on the Mississippi*, 82 (quotation); James Rudd Account Book, Filson; Martin, *Divided Mastery*.

30. Undated *Louisville Courier-Journal*, quoted in "Old Days on the River," *Waterways Journal*, September 1, 1894, 3 (quotations); James, *Antebellum Natchez*, 188–9; *Southern Literary Messenger*, 5 (February 1839): 140; *New England Magazine*, 1 (December 1831): 487–8; John T. Trowbridge, *The Desolate South 1865–1866: A Picture of Battlefields and of the Devastated Confederacy*, ed. by Gordon Carroll (Freeport, N. Y.: Books for Libraries Press, 1956), 178; John W. Reps, *Cities of the Mississippi: Nineteenth Century Images of Urban Development* (Columbia: University of Missouri Press, 1994), 109.

industry eventually centered itself in Louisville. Through 1830 workers in the city built more steamboats than Pittsburgh, and at least 750 riverboats were constructed in Louisville during the antebellum period. The city's advantages emerged early; in 1818 eight engines for steamboats were forged in Louisville at the Prentice and Bakewell Foundry, and the firm had contracts for another $70,000 worth of engines and boats. The Howard Ship Yards, located just across the Ohio River in Jeffersonville, supplied steamboat hulls for eager builders in Louisville, Cincinnati, New Orleans, Mobile, and St. Louis. Other southern cities industrialized in response to the demand for steamboats. The Wilkins, Humason and Company foundry in Natchez forged engines while two or three riverboats were built in Nashville in 1848.[31]

The net effect of these developments, from the transportation of cotton to the building of steamboats, was to strengthen and diversify the southern economy. The quantity of commerce on the western waters increased to surprising levels. In 1843 about $120,000 worth of town-to-town commerce occurred on the Mississippi and Ohio rivers, while $50,000,000 worth of goods was imported from the East, and another $50,000,000 was imported from other countries. Connections "between leading commercial, manufacturing and mercantile men" along the river were already substantial and promised to increase. When steamboats penetrated seemingly more difficult or distant markets, citizens gained access to a wider array and quantity of goods. A resident of Bowling Green, Kentucky, gaped at the huge volume of goods on the first steamboat that arrived there in 1828. Although the *United States* was a small boat, an observer wondered, "Could such an amount of goods ever be consumed?" Apparently they could, for local businessmen lobbied the Kentucky legislature to improve the river so that regular steamboat traffic was possible. Steamboats had become a reliable barometer of trade and economic vitality. Coleman Rogers, writing from Louisville, lamented that fewer steamboats on the river translated into "little bustle in our streets."[32]

---

31. Charles Preston Fishbaugh, *From Paddle Wheels to Propellers* (Indianapolis: Indiana Historical Society, 1970), 30–4; Kim M. Gruenwald, *River of Enterprise: The Commercial Origins of Regional Identity in the Ohio Valley, 1790–1850* (Bloomington: Indiana University Press, 2002), 115; Henry M'Murtrie, *Sketches of Louisville and its Environs* (Louisville: S. Penn, 1819), 132; Yater, *Two Hundred Years*, 75; James, *Antebellum Natchez*, 205; *De Bow's Review* 6 (October–November 1848): 299.

32. *Kimball and James' Business Directory for the Mississippi Valley: 1844*

The ragged march of technology brought steamboats to ascendance in the 1830s but also contributed to their decline three decades later. Railroads, at first, complemented the operation of riverboats. Short-haul feeder lines from major ports increased cargo shipped on riverboats, and both forms of transportation coordinated their schedules and increased their profits. Even after railroads began laying track parallel to major rivers, the advantages of rail transportation were not always apparent. By the 1850s, however, passengers came to favor train travel because of its relative speed and greater reliability. Shipments of freight also began to migrate to the railroads, especially along the upper Mississippi, but agricultural commodities stayed firmly wedded to steamboats on the lower Mississippi. The movement of cotton, in fact, increased until the 1880s, then began to decline and finally collapsed by 1886. By the dawn of the twentieth century, the influence of steamboats was negligible, and they had already become a curiosity of a bygone era.[33]

By the time Confederates fired on Fort Sumter, steamboats had become a key factor in southern economic and social development. They accelerated the development of the plantation economy, in turn ensuring the spread and maturation of the slave system. The boats were a catalyst for the development of the Old Southwest and accelerated regional cohesion. They helped create the Cotton Kingdom and entrenched slavery more deeply in the South. In the process, they enhanced the demand for soil in the Mississippi valley and hastened the process of Indian removal by making western settlement more desirable. Steamboats also ushered in a type of information age when they carried mail and rumors up and down the river. People realized the enormous potential of steamboats as a means to replicate "civilized" society through access to manufactured goods and luxury items. Once that happened, land west of the Appalachian Mountains seemed less remote. As Ari Kelman has noted, it seemed as if steam travel "compressed the valley's environs like an accordion, bringing the upper Ohio River and the lower

(Cincinnati: Kendall and Barnard, 1844), 2 (first quotation), 41; Helen Bartter Crocker, "Steamboats for Bowling Green: The River Politics of James Rumsey Skiles," *Filson Club Historical Quarterly* 46 (January 1972): 12 (second quotation); Coleman Rogers to Samuel Brown, January 17, 1824, Samuel Brown Papers, Filson Historical Society, Louisville, Kentucky (final quotation).

33. Hunter, *Steamboats on the Western Rivers,* 585-603; Haites, Mak, and Walton, *Western River Transportation,* 120.

Mississippi together as easily as one might fold a map, leaving Baton Rouge astride Pittsburgh." The new technology increased the profitability of cash crops and facilitated the movement of countless numbers of slaves to the region. In short, steamboats provided a measure of economic diversification even as they reinforced the hegemony of the plantation system. Steamboats did not completely reshape the region, just as the 1811 New Madrid earthquake did not obliterate all it touched, but their influence, as one contemporary noted, was obvious even to the "most dull and stupid."[34]

---

34. Kelman, *A River and its City*, 64 (first quotation); Cramer, *The Navigator*, seventh ed., 31 (final quotation).

Pits of Frustration

The Failed Transplant of British Mining Methods in Antebellum Virginia

SEAN PATRICK ADAMS

AS 1815 BEGAN, AMERICANS HAD ANY NUMBER OF REASONS TO DREAD the appearance of British ships off the East Coast. News of the February signing of the Treaty of Ghent assuaged many people's fears, but for Harry Heth's Black Heath Pits of Chesterfield County, Virginia, British vessels still carried a dangerous cargo. Heth and other colliers in the Richmond Basin hoped to capitalize upon the pent-up demand for mineral fuel after the end of Britain's wartime blockade, but the end of hostilities also meant that ships from Liverpool weighted with coal might help British merchants regain their hold on the urban markets of Boston, New York, and Philadelphia. In March, Heth's nephew, William, assessed the demand for fuel in New York City and saw an opportunity to expand the sale of the family's Black Heath coal. William implored his uncle to strike the iron while it was hot. "Coal is scarce here at this time," he wrote, as sources of foreign or domestic coal had not yet reached urban consumers in the aftermath of the British blockade. Richmond coal would earn up to twenty-five cents a bushel, William reported, a considerable increase over the prewar rate of twenty cents. By the end of the year, however, Harry's prospects had dimmed, William had left the family business, and Virginia colliers had failed to capitalize on their opportunity. By December, a Boston merchant reported to Harry that "we have been inundated with english & Scotch coal this year;" not enough high-quality coal had arrived from Virginia to change consumer preferences in urban markets.[1]

1. William Heth to Harry Heth, March 30, 1815; Thomas B. Main to Harry Heth,

Mistiming the shipment of goods to market was an everyday occurrence in the economy of the early United States, and perhaps Heth's failure to exploit the situation in 1815 offered only a temporary setback to colliers of the Richmond Basin. After all, Virginia colliers entered the first decades of independence expecting to cultivate a national market and had every reason to be optimistic. As cities on the eastern seaboard grew, local reserves of firewood declined, and the demand for coal as a source of heating fuel increased. Blacksmiths, bakers, and small manufacturers, moreover, found coal to be essential to their continued prosperity. Previous generations had relied upon British imports when they needed mineral fuel, but American policymakers looked to domestic coalfields to provide a measure of self-sufficiency in mineral fuel following the Revolution. In 1787, Thomas Jefferson described the Richmond area as "replete with mineral coal of a very excellent quality" and noted that a number of proprietors opened coal pits which, "before the interruption of our commerce [,] were worked to an extent equal to the demand." In his now-famous *Official Report on Publick Credit*, Alexander Hamilton marked Virginia coal as an important domestic and industrial fuel. Political economist and industrial enthusiast Tench Coxe expected the collieries along the James River to provide inexpensive coal so that British imports "will be rendered a very losing commodity." He also hoped that the coastwise trade would act as a "valuable nursery" for American seamen. Those accounts all predicted that the Old Dominion could provide mineral fuel for the nation's growing economy for decades to come; they also suggested that the former American colonies could use their ample reserves of coal to emulate British economic success and produce their own dynamic manufacturing sector. Thus Virginia would serve as the young nation's Newcastle—a region that could ship coal across the American coast to provide cheap fuel for manufacturing, warmth in the cold winter months, and self-sufficiency in energy.[2]

A closer examination, however, reveals that Heth's missed opportunity in the winter of 1815 was no aberration. It represented

---

December 2, 1815, Heth Papers, Alderman Library, University of Virginia, Charlottesville, Virginia, hereinafter UVA.

2. Thomas Jefferson, *Notes on the State of Virginia* (Chapel Hill: University of North Carolina Press, 1955 [1787]), 28; Alexander Hamilton, *Official Report on Publick Credit* (Washington: Government Printing Office, 1790), 249–50; Tench Coxe, *A View of the United States of America* (Philadelphia: William Hall and Wrigley and Berriman, 1794), 180–81.

in miniature the history of disappointment in the Richmond Coal Basin and the Virginia collier's difficulties in copying the success of his colleagues in Great Britain. By raising coal cheaply and effectively, British colliers served as a keystone in their nation's burgeoning industrial economy. Harry Heth and other mine owners in the Richmond Basin aspired to unleash the power of mineral fuel in the United States in a similar fashion. In some respects, the Richmond Basin succeeded. It accounted for over half of seaboard coal consumption in the United States through 1826 and continued to play a significant role in fuel markets through the 1830s. But in other ways, the Richmond Basin failed to capitalize on its many advantages. Colliers there implemented some sophisticated methods of shaft construction, ventilation, and coal extraction that had been perfected in British mines. They struggled, however, to construct other elements of a successful coal trade such as the effective use of steam engine technology, efficient transport systems, and the coordination of production. Despite local colliers' longstanding interest in developing the region, various natural advantages, and the ready knowledge of British coal-mining methods, the Richmond Basin's prospects were never fully realized.

This failure of national proportions had local origins. Why were local conditions in Virginia inhospitable for such a straightforward activity such as the extraction of coal? Slavery, perhaps the obvious answer in most case studies of southern industrialization, provides only a partial explanation. On the one hand, the dependence upon slave labor certainly complicated coal mining in the Richmond Basin, just as it hindered the proliferation of iron furnaces, textile mills, and railroads in the antebellum South. On the other hand, the use of slave labor did not create a major impediment to the expansion of the coal trade, as colliers of the Richmond Basin effectively blended free and bonded labor in their mines. Slaves certainly could raise coal effectively, but the broader elements of agrarian political economy in the Old Dominion, of which slave labor undoubtedly played a major role, frustrated colliers of the Richmond Basin when they sought to expand their trade. Even if they could raise coal to the mouth of the mine in the short term as effectively as their national and international colleagues, Virginia colliers faced great hurdles in the long term. Operating on the margins of a plantation economy denied them the extensive economic and political support a fledgling trade required, and by the time mineral fuel became essential to American industry at the onset of the Civil War, the great potential of the Richmond Basin was a distant memory. Anthracite coal, with extensive support from Pennsylvania's entrepreneurial community, state government, and local boosters, had stormed urban fuel markets by the 1830s. In this regard, the story of the Richmond Basin provides an early perspective on the difficulties southern industrialization faced later in the antebellum period. Slavery never hindered the actual mining of coal, but it did chip away at its competitive edge. Heth and other colliers believed that they could create an "American Newcastle" in Richmond. The British system, however, sank only shallow roots in the Virginia soil, and Richmond's "coal pits" modeled on British methods became pits of frustration in the Old Dominion.[3]

To modern eyes, the Richmond Coal Basin appears quite modest. But early American nationalists, desirous of their own sources of Britain's "black diamonds," viewed the field as priceless. This triangular coalfield sits tantalizingly close to Richmond's deepwater port at Rockets. The James River bisects the Richmond basin, and the 150-square-mile coalfield is crisscrossed by a number of small creeks. By the early eighteenth century, local residents had found coal outcroppings along these ubiquitous creek beds. Because the coal is bituminous in rank, it burns rather easily, and despite the dense smoke associated with bituminous coal, it served well in blacksmith forges, ovens, and other small-scale manufacturing concerns. When the growth of colonial cities increased the demand for mineral fuel, Virginia's early colliers worked these exposed seams using crude methods of extraction such as the open-air "bell pit" used in easily accessed coal deposits in England. On the eve of the American Revolution, Richmond's deep black coal dominated the small, but growing, coastwise trade.[4]

3. There is a large body of literature on the failure of southern states to keep up with their northern counterparts in terms of antebellum industrial growth. Some representative works include Eugene Genovese, *The Political Economy of Slavery: Studies in the Economy and Society of the Slave South* (New York: Random House, 1965); Gavin Wright, *The Political Economy of Slavery* (New York: W. W. Norton, 1978); Fred Bateman and Thomas Weiss, *A Deplorable Scarcity: The Failure of Industrialization in the Slave Economy* (Chapel Hill: University of North Carolina Press, 1981); John Majewski, *A House Dividing: Economic Development in Pennsylvania and Virginia before the Civil War* (New York: Cambridge University Press, 2000).

4. Hugh Jones, *The Present State of Virginia: From Whence is Inferred a Short View of Maryland and North Carolina* (Chapel Hill: University of North Carolina Press, 1956 [1724]), 144; Kathleen Bruce, *Virginia Iron Manufacture in the Slave Era* (New York: Century Company, 1931), 88.

Since most early Americans relied upon coal from Newcastle and other British sources, the Richmond Basin's potential remained virtually untapped during the Revolutionary era. Familiarity and the imperatives of empire led colonial consumers to import Newcastle coal across the Atlantic. In 1770, for example, Britain exported over 8,700 tons of coal to the American colonies. Although wood fuel still dominated American hearths and furnaces, Britain's successful use of mineral coal did have an impact in its colonies. In 1773, for example, Benjamin Franklin modified the design of his famous stove to make it "fit only for burning Pit-Coal." Not surprisingly, British imports reappeared in American ports following the Treaty of Paris, much to the chagrin of Virginians. One Richmond resident blamed British imports for the inability of some local entrepreneurs to "lay out their capitals" in his local coalfield.[5]

In 1798, twenty-one Richmond colliers petitioned Congress for a higher tariff, arguing that their coal could supply "for centuries, the demand of the United States" and asked that higher tariffs "convert this circumstance to the accomplishment of national advantages." Congress denied that particular remonstrance but eventually helped discourage the expansion of coal imports to the United States by revisions in the tariff structure. In 1808, British coal went for about eighteen cents per bushel at Liverpool and paid a duty of five cents. When merchants figured in the cost of freight, British coal had to sell for at least thirty cents a bushel to net any profit, making it uncompetitive with Virginia coal that sold at about twenty cents to the bushel in Philadelphia.[6] Hostility toward British imports, the desire to encourage domestic production, and the need for federal revenues inspired Congress to increase tariff levels to ten cents a bushel in 1812—about 15 percent of the wholesale price of British coal.[7]

5. Benjamin Franklin to Jacque Barbe-Dubourg, June 29, 1773, in *The Papers of Benjamin Franklin*, vol. 20, *January 1 to December 31, 1773*, ed. William Willcox (New Haven: Yale University Press, 1976), 251; Thomas Mann Randolph to Thomas Jefferson, May 3, 1790, in *The Papers of Thomas Jefferson*, vol. 16, *November 30, 1789, to July 4, 1790*, ed. Julian Boyd (Princeton: Princeton University Press, 1961), 410.

6. See the report of Joseph Gilpen on a proposed Chesapeake and Delaware canal. Albert Gallatin, "Report on Roads and Canals," in *American State Papers, Miscellaneous*, vol. 1, *Documents, Legislative and Executive of the Congress of the United States* (Washington: Gales & Seaton, 1834), 760.

7. Price calculation is based on Arthur Cole's calculations of the wholesale price of coal in New York City in 1812, which fluctuated between $27 and $28 a chaldron. See Arthur Cole, *Wholesale Commodity Prices in the United States, 1700–1861* (Cambridge: Harvard University Press, 1938), 164.

Combined with the precipitous decline in all foreign imports during the War of 1812, this relatively high tariff provided a window of opportunity for the American coal trade. Through the 1840s, duties on foreign coal imports ranged from 10 to 25 percent of the price of British coal in large markets like New York City. Although not explicitly designed to be protective, the tariffs effectively disadvantaged a high-bulk, low-value commodity like coal in American seaports. In the decades after 1815, colliers' persistent cries for higher protective duties sounded increasingly shrill as British imports never again exceeded 10 percent of American consumption.[8]

The twenty-five or so coal-mining operations in the Richmond Basin found themselves in an important strategic position when coastwise markets for fuel opened up in the postwar United States. Higher duties and the growth of small manufacturing in American cities gave Richmond colliers the opening they had sought for years. They could not bury Boston, Philadelphia, and New York under a blizzard of coal, as Virginia accounted for about one-fifth of American production in 1815 and trailed the leading coal-producing state, Pennsylvania, by a large margin. But Pennsylvania's bituminous coal lay in its western counties, far outside striking distance of coastal cities, and its anthracite reserves remained unexplored and, as yet, untapped. As the only sizeable coalfield within easy reach of coastwise vessels, the Richmond Basin offered the cheapest alternative to British imports to dominate the seaboard trade. The "first mover" advantage, as well as the attention of federal officials like Albert Gallatin, marked the Richmond Basin as the early center of the American coal trade.[9]

As the leading coal producer in the region, Harry Heth's Black Heath Coal Pits set the pace for other producers in the Richmond Coal Basin. The son of an English immigrant, Heth cultivated contacts with British colliers at the same time that he lived and

8. "Duty on Coal," in *American State Papers*, vol. 5, *Documents, Legislative and Executive of the Congress of the United States* (Washington: Gates & Seaton, 1832), 553; F.W. Taussig, *The Tariff History of the United States*, 8th ed. (New York: G. P. Putnam's Sons, 1931), 17; Arthur Cole, *Wholesale Commodity Prices, 1700–1861* (Cambridge: Harvard University Press, 1938), 178; Howard Eavenson, *The First Century and a Quarter of American Coal Industry* (Pittsburgh: Privately Printed, 1942), 12; Curtis P. Nettels, *The Emergence of a National Economy, 1775–1815* (New York: Holt, Rinehart, & Winston, 1962), 324–35.

9. Rhode Island had a small anthracite coal field but never effectively competed with British, Virginia, or Pennsylvania coal during the Early Republic. For antebellum production statistics on coal, see Eavenson, *American Coal Industry*, 426–34.

socialized among his fellow planters in the Old Dominion. He owned a large plantation on the outskirts of Richmond and was as wedded to the institution of slavery as any tobacco planter in the region. But Heth also served as a model of entrepreneurship among local colliers, and the Black Heath Pits were the largest single shipper of coal to market for the first two decades of the nineteenth century. The Heth name became synonymous with the Richmond Coal Basin, as John Heth took over the business after Harry's death in 1821 and expanded both the scale and reputation of the Black Heath Pits. In fact, when the U.S. Navy advertised for twenty-five thousand bushels of bituminous coal in 1837, they established Black Heath coal as their quality standard. Two years later, John Heth's operations produced up to twenty-five hundred bushels of coal a day and claimed an annual capacity of two million bushels. As the leading colliers in the region at the time of its greatest promise, the Heth family drew upon both personal experience and trans-Atlantic contacts to adapt the techniques used in British mines to the particular contours of eastern Virginia's coalfields. They set the standard in the Richmond Basin, but their blueprint for success had been perfected overseas.[10]

Virginia colliers realized, for example, that a blend of skilled managers and a variety of underground and aboveground workers provided the foundation for the British coal trade, and the Virginians sought to replicate this system in the Richmond Basin. The experienced "hewers" oversaw the extraction of coal at the face of the seam and drew upon decades of hard-won mining wisdom. A profitable mine also required workers to brace the underground shafts and galleries and tend to the machines needed for lifting coal and pumping water out of the mine. The job of cutting, loading, and hauling the coal required less experience but plenty of brute force. Since these jobs were unpleasant and dangerous, British colliers used a system of "bonds" in order to retain their labor force. These contracts defined the rate of payment, obligated the miner's family to keep company housing in good shape, and set up an arbitration process for the resolution of disputes. As skilled workers were in high demand among British colliers, two crucial components of the

10. Bruce, *Virginia Iron Manufacture*, 101–2; Ida J. Lee, "The Heth Family," *Virginia Magazine of History and Biography* 42 (1934): 277; Ronald Lewis, *Coal, Iron, and Slaves: Industrial Slavery in Maryland and Virginia, 1715–1865* (Westport, CT: Greenwood Press, 1979), 52, 64–65.

bond was its annual length and its no-strike clause, both of which could be enforced by local law officials. Although it sounded quite restrictive, many mine workers favored the bond system, especially when they could force mine operators to come up with "binding money" to induce them to sign during a particularly tight labor market. In controlling workers and stunting labor militancy, British mine operators eventually began requiring a "discharge note" from new employees as proof that they had left their last jobs in good standing. By the early nineteenth century, "binding" had evolved into a standard practice in the labor relations of the coal industry much like the lease agreements between landowners and mine operators. With this stable industrial community in place, British managers could implement scientific methods of mining and new technology in their coalfields without fear of disruption or bankruptcy.[11]

The degree of control over the laborer in the British fields suggests that slave labor might have adapted well to coal mining, but the predominance of slavery in Virginia made the exact replication of British methods of training and securing labor difficult in the Richmond Basin. With a population of less than ten thousand in 1810, Richmond could not equal the employment opportunities for unskilled laborers offered by larger cities like New York, Philadelphia, or Baltimore. The use of slaves for domestic and craft labor restricted the number of available wage workers in Richmond even further. Virginia colliers therefore relied upon a limited surplus of local labor, which often included whites and free African Americans, as well as enslaved workers. Such combinations of free and slave labor could function, as prior historians of Virginia's early coal and iron industries have well demonstrated, but in other ways Virginia's economy hamstrung colliers' efforts to secure a workforce.[12]

11. Scottish miners tended to enter into longer bonds than mine workers in England. In fact, up until the early nineteenth century, some Scottish miners were bound for life to a single employer. Michael W. Flinn, with David Stoker, *The History of the British Coal Industry*, vol. 2, *1700–1830: The Industrial Revolution* (Oxford: Clarendon Press, 1984), 329-66; J. H. Morris and L. J. Williams, "The Discharge Note in the South Wales Coal Industry, 1841–1898," *The Economic History Review* (New Series) 10 (1957): 286–93.

12. For more on the use of skilled labor in Virginia's mines, see Robert Starobin, *Industrial Slavery in the Old South* (New York: Oxford University Press, 1970), 146–89; and Lewis, *Coal, Iron, and Slaves*, 179–209. In a study of the early American iron industry, John Bezís-Selfa suggested that scholars of early industry should "abandon their tendency to examine free and slave labor in isolation from one another" ("A Tale of Two Ironworks: Slavery, Free Labor, Work, and Resistance in the Early Republic," *William and Mary Quarterly* 3rd. ser., 56 [1999]: 700).

In the Richmond Basin, for example, agricultural work always took precedence over mining in the allocation of labor. The inherent hazards of coal mining, such as fires, shaft collapses, and floods, made many owners reluctant to hire out their slaves for such dangerous work. One mine explosion in 1839, for example, killed forty-five black miners along with two white overseers. Virginia colliers used their own slaves in mines but still faced severe labor shortages. Harry Heth owned forty-one slaves at the time of his death but needed between fifty and one hundred to operate his plantation and coal mines. Some of the slaves owned by colliers became quite skilled in the art of mining, but a great deal of the work done in the coal pits simply required brute force. In fact, the backbreaking work—hacking at coal seams with picks, lugging loads of coal through mucky tunnels, propping up sagging ceilings with timber— meant that healthy male slaves would have been preferred by colliers. Since agricultural labor also put a high premium on this type of slave, rounding out a sufficient labor force in the coal mines remained a constant problem. In 1805, Mantapike planter Richard Brooke admonished Heth for requesting hired slaves for his coal pits during harvest. In 1812, another planter refused to replace a slave who had fled Heth's coal mines and returned home. "I have been trying to hire hands for you ever since I saw you & have not as yet procured a single one," one friend wrote to Heth in 1819, "& I am afraid I shant be able to get one single one in the neighborhood." Implementing slave labor in coal mining thus raised the expense and multiplied the hassles of raising coal in the Richmond Basin.[13]

Early miners dug open-air quarries along the James River, but as shafts pursued the coal seams deeper into the earth, mining operations required knowledgeable workers. Slaves could eventually learn mining methods, but this most likely entailed years of hands-on training. Harry Heth worked slaves at every skill level in his

13. Richard Brooke to Harry Heth, June 7, 1805, Harry Randolph to Heth, June 1, 1812; H. B. Christian to Heth, March 1, 1819; Heth Papers, UVA; Lewis, *Coal, Iron and Slaves*, 66. Recent research on Virginia suggests that either free or slave labor could be successfully adapted to industrial pursuits, so the problem facing Richmond colliers should be viewed as one of scarcity, not necessarily of quality. See Charles Dew, *Bond of Iron: Master and Slave at Buffalo Forge* (New York: Norton, 1994); Midori Takagi, *"Rearing Wolves to Our Own Destruction": Slavery in Richmond, Virginia, 1782–1865* (Charlottesville: University of Virginia Press, 1999); John Bezís-Selfa, *Forging America: Ironworkers, Adventurers, and the Industrious Revolution* (Ithaca: Cornell University Press, 2004).

mines and positioned them in areas of high responsibility. Slave miners needed to be dependable, as Virginia mines used the complicated British "pillar-and-breast" system in deeper works. This system requires large blocks of coal—the "pillars"—to support the roof of the mine while miners dig the coal from small rooms, or "breasts." Mine foremen needed to know how much coal to leave standing in order to avoid disastrous cave-ins. Deadly methane gas also seeped from coal seams in the Richmond Basin. If allowed to accumulate, it could ignite in a deadly fireball or smother the miners. Skilled mine foremen therefore commanded the highest wages; if they were slaves, they received more incentives to compensate for their risky position. Some Virginia colliers hired recent immigrants from the other side of the Atlantic in order to bring in skilled labor. By 1818, for example, Harry Heth employed two Scots in the Black Heath mines in hopes of increasing production. Although these imported miners could "win" more coal (as miners called the process), they also could implement the most modern methods of ventilation and roof support used in British mines.[14]

How easily could coal-mining technology travel with immigrants? Case studies in other industries suggest that it could move with little difficulty. In fact, the early histories of the glass and iron industries demonstrate that industrial technology traveled across the Atlantic despite significant cultural and linguistic barriers. The German immigrant Caspar Wistar was able to transplant glassmaking techniques from the Palatine to colonial New Jersey. By 1741, Wistar's United Glass Company was valued at £1,000 and employed glassmakers of both German and American origin. Henry William Stiegal, who migrated from Cologne to Pennsylvania in 1750, brought along skills in iron smelting, casting stove backs, and making glassware. Perhaps more significant, although more difficult to trace, are the more mundane transfers of technology that occurred among less prominent immigrants. John Bezís-Selfa argues that eighteenth-century iron adventurers actively recruited Germans both locally and from abroad in order to take advantage of their familiarity with ironworking technology. In exchange, many of these new arrivals used their prestigious position on the shop floor

14. John Grammer, "Account of Coal Mines in the Vicinity of Richmond, Virginia," *American Journal of Science* 1 (1819): 125–30; Bruce, *Virginia Iron Manufacture*, 95–96; Ronald Lewis, *Black Coal Miners in America: Race, Class, and Community Conflict: 1780–1980* (Lexington: University Press of Kentucky, 1987), 4–5.

to develop their own social and business contacts. The early American iron trade bore a heavy German imprint, and immigrants continued to serve in both skilled and unskilled positions for decades.[15]

The transfer of coal-mining technology followed this general pattern, as British colliers drew international attention for their mining methods, and British migrants brought this knowledge with them to America. Late-eighteenth- and early-nineteenth-century travelers often marveled at the ingenious nature of British coal mines, and American accounts of the mother country's coal industry, in particular, brimmed with admiration. Jonathan Williams, accompanying Benjamin Franklin on a tour of northern England in 1771, remarked on how the efficient system of canals and cranes "manage this great Work in a very expeditious & cheap Manner" in the Duke of Bridgewater's coal mines. During visits to various colliers in 1801, the young American Thomas Peters Smith carefully noted the steam engines used to lift both miners and coal and the application of railroad technology in the mines. Once in America, emigrants advertised their familiarity with the world's leading coal trade. "A man from the Coal Mines in the north of England, who is well acquainted with boring for, and raising of Coal," for example, announced that he was looking for employment in the Philadelphia area in 1803. After the War of 1812 ended, these migrants made their way to the Richmond Basin. In addition to the anecdotal evidence of Heth's hire of two Scottish miners, census data suggests that foreigners could have accounted for up to 13 percent of the manufacturing labor force in Heth's Chesterfield County by 1820. Richmond Basin colliers drew upon immigrants to implement British mining methods, such as the aforementioned breast-and-pillar method of coal extraction, as well as the technical knowledge needed for reinforcing walls and ceilings of mine shafts, ventilating work areas, and

15. Rosalind Beiler, "Peterstal and Wistarburg: The Transfer and Adoption of Business Strategies in Eighteenth-Century Glassmaking," *Business and Economic History* 26 (1997): 343–53; Frederick William Hunter, *Stiegel Glass* (New York: Dover, 1950), 5-60; Bezís-Selfa, *Forging America*, 120–24; Charles Steffan, "The Pre-Industrial Iron Worker: Northampton Iron Works, 1780–1820," *Labor History* 20 (1979): 89–110; Thomas M. Doerflinger, "Rural Capitalism in Iron Country: Staffing a Forest Factory, 1808–1815," *William and Mary Quarterly* 59 (2002): 3–38. For more on the early organization of Pennsylvania's iron industry, see Paul F. Paskoff, *Industrial Evolution: Organization, Structure, and Growth of the Pennsylvania Iron Industry, 1750–1860* (Baltimore: Johns Hopkins University Press, 1983).

dealing with troublesome methane leaks. Such confidence seemed justified. One mining expert, William Keating, praised the "active and enterprising character" of the English miner in 1821, "in whom we find a happy mixture of enterprise and perseverance, of enthusiasm in undertaking, and skill in accomplishing, his operations."[16]

Mine drainage, however, posed a considerable challenge that required a more extensive technical solution than the average British miner could provide. As the coal "ran" deeper into the earth, colliers used "drifts" to intersect the coal at an angle, or "shafts" to reach the seams when they ran parallel to the surface. As the drifts and shafts ran deeper in the ground, Virginia miners of the Early Republic sometimes hit abandoned works that had filled with water. In 1801, Samuel Paine reported on the difficulty of removing the stagnant water from older mines. If his mine could only avoid this setback, he wrote to his business partner, it could produce enough coal to turn a nice profit. Rainwater also ran into open shafts, which compounded the difficulties of working underground. As if gas explosions, suffocation, and collapsed tunnels were not enough to make miners miserable, the flooding of mine shafts added a dank and slushy working environment to the equation.[17]

The Black Heath Pits had two shafts descending six hundred and seven hundred feet deep by the 1820s—the deepest in the United States at the time. But waterlogged mines remained one of the area's most persistent tribulations. British miners applied steam technology to this problem in 1712 with the advent of the Newcomen pump, named after inventor and Dartmouth ironmonger Thomas Newcomen. This steam-powered vacuum pump revolutionized coal mining by allowing miners to dig pits hundreds of feet

16. Quotes on British mining methods come from the entry of May 21, 1771, *Journal of Jonathan Williams Jr., 1771–1772*, Jonathan Williams Papers, and the entries dated November 11, 23, 25 and December 18, 1801 in Thomas Peters Smith Journals, vol. 4, American Philosophical Society, Philadelphia, PA, hereinafter APS. Although the 65 resident aliens listed in 1820 were only a drop in the bucket compared to Chesterfield's 7,543 free and 9,513 enslaved residents, they most likely were a strong presence in the region's 501 persons engaged in manufacturing. United States Census Office, *Census for 1820* (Washington: Gales & Seaton, 1821), 106; *Philadelphia American Daily Advertiser*, February 23, 1803; William Keating, *Considerations upon the Art of Mining. To Which Are Added, Reflections on Its Actual State in Europe, and the Advantages Which Would Result from an Introduction of This Art into the United States* (Philadelphia: M. Carey & Sons, 1821), 72.

17. Samuel Paine to Richard Morris, October 27, 1801, Morris Family Papers, Alderman Library, UVA.

deep and still remove water from the workings at a lower cost than with previous systems. In 1744, Griff colliery in Warwickshire estimated that a Newcomen pump cost about £150 in fuel, maintenance, and labor, which replaced the £900 cost in feed and labor needed for the team of fifty horses previously used to power water pumps. As improvements to the original Newcomen design, such as the one devised by the firm of Boulton and Watt in 1769, came into use, cost savings from steam-powered pumps allowed mines to delve into seams of coal previously thought unreachable. William Keating urged the "constant application of machinery wherever it is practicable" in American mines. Successful application of steam technology to mining practices thus emerged as an indispensable aspect of the British style of mining and an option that Virginia colliers would have enthusiastically embraced.[18]

Although adopting British steam technology seemed like a good idea to Virginia's colliers, major hurdles stood in the way. Copper mines in Connecticut and New Jersey used Newcomen engines to clear water from their workings throughout the eighteenth century, but extensive knowledge of those systems never extended beyond those particular regions. Indigenous sources of steam power never materialized in the Old Dominion. Here Heth's hiring of two Scottish immigrants in the period after the War of 1812 looms large, as immigrants could bring both theoretical and practical knowledge of complex technological systems with them to the United States. A historian of intellectual piracy, Doron Ben-Atar, argues that enticing skilled laborers to emigrate across the Atlantic provided the bulk of technology transfer during the Early Republic. So in addition to well-established practices in coal mining, perhaps immigrants would bring cutting-edge British technology to the Richmond Basin. "Successful industrialization," Ben-Atar claims, "depended on manufacturers' ability to pirate British technology primarily through enticing skilled British artisans to emigrate."[19]

Again, other industries point toward a successful transfer. Textile manufacturing methods, despite being jealously guarded by British

18. For more on the history of the Newcomen pump and its successors, see Flinn, *History of the British Coal Industry*, 114–28; Keating, *Considerations upon the Art of Mining*, 82.

19. Louis Hunter, *A History of Industrial Power in the United States, 1780–1930*, vol. 2, *Steam Power* (Charlottesville: University of Virginia Press, 1985), 3–4; Doron S. Ben-Atar, *Trade Secrets: Intellectual Piracy and the Origins of American Industrial Power* (New Haven: Yale University Press, 2004), 152.

authorities, made their way across the Atlantic during the 1790s and early 1800s. Although the most famous example of this was found in Samuel Slater's Pawtucket factory, skilled immigrants brought modern textile manufacturing with them to the Philadelphia region and other areas of New England. "Clearly, without technology diffusion," historian David Jeremy has argued, "industrialization is long delayed or never takes place." Replications of British textile mills dotted the American countryside by the 1820s, despite British authorities' efforts to halt the emigration of skilled workers and ideas. Although the adoption of British methods required significant modification in order to suit American conditions, the gradual transferal of skills from newly arrived artisans, industrial espionage, and the successful modification of British textile technology allowed American textile manufacturing to grow substantially. Even the most commonly cited example of endogenous southern technology, the saw gin for cleaning cotton, came from a recent emigrant from New England to the South, Eli Whitney. Whitney's initial attempts to protect his invention failed, moreover, as mechanics copied or modified his design without reprisal. So from the raw cotton to the finished product, American textiles relied upon exogenous sources of technology and owed a great debt to the rapid dissemination of technical information in the Atlantic World. Even complex industrial systems can be replicated with time, effort, and a favorable environment.[20]

The success in implementing steam technology in the Richmond Basin, however, proved more elusive than the adoption of more basic mining techniques from overseas, and the presence of slavery in Virginia seemed to complicate the use of steam engines in the Richmond Basin. Harry Heth first attempted to contract with local engineers to construct steam engines for this purpose in 1810, but these efforts proved ineffectual.[21] Heth then contacted Oliver

20. David Jeremy, "British Textile Technology Transmission to the United States: The Philadelphia Region Experience, 1770–1820," *Business History Review* 47 (spring 1973): 24–52; quotation from Jeremy is from *Transatlantic Industrial Revolution: The Diffusion of Textile Technologies between Britain and America, 1790–1830s* (Cambridge: MIT Press, 1981), 262; David Gwyn, "Tredegar, Newcastle, Baltimore: The Swivel Truck as a Paradigm of Technology Transfer," *Technology and Culture* 45 (2004): 778–94; Angela Lakwete, *Inventing the Cotton Gin: Machine and Myth in Antebellum America* (Baltimore: Johns Hopkins University Press, 2003), 47–71.

21. George Easterly of Richmond offered to construct a system to raise water but wanted a $1,000 advance plus an estimated $250 to $300 in wages. Daniel French, a

Evans, the noted steam engine manufacturer of Philadelphia, in 1813 to inquire about sending someone to design and implement a steam-powered system for pumping water from his mines. Evans agreed to build him an engine but refused to send any of his employees to Virginia:

> The workmen here have embraced a prejudicial idea of the customs of your country, they think that if a master mechanic goes into your employ, and will refuse to work a task himself but keep himself clean and talk big and help himself freely to brandy, wine &c, that you will treat him as a gentleman. But if he lays his own hands vigorously to the work (which he will be compelled to do because slaves cannot do it) you immediately rank him with slaves, with whom he is forced to work, and will not think him entitled to more than half the wages of a gentleman mechanic, who does not earn a cent by his hand.

Evans dismissed the notion that his steam engine could operate with slave labor. "I fear [you] have wrong ideas if you think slaves can keep a steam engine in order," he wrote. "A man must be free before his mind will expand so much." Apparently the pumping engine of Evans's design failed to yield results without the expertise of a northern engineer to keep it running. "If you had performed with it not only you but all the pits about Richmond might have been receiving the benefits of my invention," Evans lamented in 1815, "and I might have had the profits of making the Engine." Despite being well aware of steam's power to raise water, Virginia colliers found that their position in a slave-based plantation society complicated the transfer of advanced pumping systems to their mines.[22]

The delay in implementing steam-pump technology for mine drainage made some of the typical problems of industrial slavery much worse. Hiring out slaves was a fixed cost, so the profit margin

New Yorker living in Richmond, entered into an agreement with Heth to pump water from the Black Heath pits for $5,000. Apparently both of these efforts had failed by 1813. George Easterly to Harry Heth, June 3, 1810; Agreement between Daniel French and Harry Heth, November 30, 1811; Harry Heth to Thomas Taylor, January 31, 1813, Harry Heth to David Meade Randolph, June 22, 1814, Heth Papers, UVA.

22. Oliver Evans to Harry Heth, June 26, 1813; July 14, 1813, Heth Papers, UVA; Oliver Evans to Harry Heth, June 20, 1815, Heth Family Papers, Virginia Historical Society, Richmond, Virginia, hereinafter VHS; D. S. L. Cardwell, "Power Technologies and the Advance of Science, 1700–1825," *Technology and Culture* 6 (1965): 188–207.

of a day's mining sagged noticeably if miners were forced to do anything other than raise coal. Miners refer to activities such as bracing the ceilings of tunnels and galleries, removing water and rock, and managing the ventilation of the mine as "dead work." Waterlogged mines, in particular, consumed hundreds of lost hours. In later decades, "dead work" was the responsibility of the miner, who was paid for the amount of coal raised and not by the hour. But for Virginia colliers, who hired labor at a fixed rate, "dead work" sliced profit margins even more. "At the Christmas holidays when the hired negroes went home and before we hired others, the whole of those workings were filled with water," one miner reported. "It required constant work for ten days to get it out. Afterwards we could keep it under in two hours every day or less." Simply put, workers who raised more water than coal lost money for Virginia's colliers.[23]

Without a local, or even national, technological community to rely upon, Virginia colliers again sought help from overseas. In 1814, Heth wrote to a friend in London that "This d___d war has all but ruined me," but apparently the hostilities could not keep him from inquiring about a British engineer for his mines. After delays, unanticipated expenses, and other setbacks, Heth secured a steam engine pumping system through the British firm of Boulton and Watt. One manager at Black Heath predicted that the new steam engines "will afford a vast quantity of the most elegant coal," the charges for shipping machinery from distant manufactories and the high cost of training skilled mechanics to maintain steam engines made British technology particularly expensive to implement in Richmond Basin mines. Heth was not confident in its ability to cut costs enough to warrant his seven thousand dollar expenditure. "The Engineers who are putting it up, appear sanguine," he fretted in one letter to a business contact, "I hope they may be right." The Black Heath Pits did see a significant upsurge in production, as Heth's operations raised 567,245 bushels of coal in 1817—a significant change from the wartime low of 45,867 bushels in 1814—but could not reach the prewar high of 830,742 bushels set in 1812. Production thus increased, but pumping water from the Black Heath Pits remained an expensive endeavor throughout their existence.[24]

23. Lewis, *Coal, Iron, and Slaves*, 57; Priscilla Long, *Where the Sun Never Shines: A History of America's Bloody Coal Industry* (New York: Paragon House, 1989), 38.

24. Grammer, "Account of the Coal Mines in the Vicinity of Richmond," *American Journal of Science* 1 (1819): 127; Henry Heth to Harry Heth, February 24,

Even if colliers found sufficient labor and successfully implemented the latest technology, they faced an ineffective transportation network between the Richmond Basin and urban markets. During the eighteenth and early nineteenth centuries, many colliers in eastern Virginia used turnpikes to transport their coal. As traffic swelled during the War of 1812, however, local turnpike operators found that the higher toll revenue hardly covered the increased wear and tear on their roads caused by hauling coal. The managers of Richmond-area turnpikes often asked the legislature for the authority to raise the rates on coal wagons. Ton-mile rates in the early nineteenth century ran about thirty to seventy cents, which quickly added to the cost of coal, which averaged about three or four dollars a ton. So even though the Richmond area witnessed a dramatic increase in turnpike chartering following the War of 1812, these roads were of limited use to colliers.[25]

Here again, Richmond colliers might have looked toward Britain and its successful use of waterways to transport coal in the eighteenth-century trade. Since the overland transport of coal doubled its price in ten miles in England, colliers used waterways whenever possible. To streamline water transport, colliers implemented a system of riverside quays, called "staiths," to transfer coal from mine wagons to vessels, or "keels." American visitors often reported upon the advantages of this system for transporting bulk commodities. In 1771, Jonathan Williams found that the Duke of Bridgewater's canal "brings out his Coals in a very easy Manner." Convenience also led to cost efficiency, as historians estimate that waterways could carry coal to market twenty times farther than land transport at the same unit cost.[26]

The Richmond Basin's system paled in comparison to the British approach to transporting coal. Colliers in Virginia stockpiled their

1818; Harry Heth to Thomas Railey and Brother, August 11, 1816, Heth Papers, UVA; Howard Eavenson, "Some Side-Lights on Early Virginia Coal Mining," *Virginia Magazine of History and Biography* 5 (1942): 203–4; Starobin, *Industrial Slavery in the Old South*, 146–89; Lewis, *Coal, Iron, and Slaves*, 179–209.

25. Legislative Petitions from Richmond City, December 13, 1813, and November 9, 1814. Legislative Petition Collection, LVA; *Richmond Enquirer*, March 2, 1816; George Rogers Taylor, *The Transportation Revolution, 1815–1860* (New York: Holt, Rinehart, & Winston, 1951), 133.

26. Entry of May 21, 1771, Williams Journal, APS; John Holland, *The History and Description of Fossil Fuel, the Collieries, and Coal Trade of Great Britain* (London: Whittaker & Co., 1835), 348–53; Flinn, *The History of the British Coal Industry*, 146, 163–71, 180–89.

product at the banks of the river and eventually dumped it into waiting barges. After the barges floated downstream to Richmond, workers shoveled the coal into wheelbarrows and carried it to a series of warehouses near the deepwater port at Rockets. When coastwise vessels were ready to take the coal to Philadelphia, New York, and Boston, workers once again shoveled the coal into wheelbarrows, took them to the ships, and dumped the coal into a ship's cargo hold. After days of rough handling, the bituminous coal crumbled into less-combustible pieces. The lack of an integrated system of staiths and keels delayed the movement of coal to market and reduced its value. Harry Heth fielded a number of complaints about the quality of his coal. In 1811 J. P. Pleasants of Baltimore wrote that "the quality of the Coal lately recd had been so indifferent that I have lost the sale of a great deal. It is not one, two or three who complain, but all." One Boston merchant, Thomas Main, complained to Heth in December of 1815 that his last shipment was "small and dirty" and must have been the "scrapings of the yard." Main decided to cancel any orders for the spring and summer of next year, and Heth lost a valuable contract. Chesterfield County colliers complained to the legislature that the existing system delayed and damaged their shipments as they worked through the various locks and transfers. "[T]he Quality of the coal is so naturally injured," miners argued, "that it can never gain a sufficient character in the northern markets to offer an inducement to us to use [the canal]."[27]

Political circumstances related to Virginia's slave-based political economy exacerbated the failure of the Richmond Basin's transportation network to facilitate the coal trade. Colliers noted, for example, that the James River Company kept water levels high enough to facilitate the moving of lighter commodities such as tobacco, wheat, and corn. But coal barges designed to carry 180 to 240 bushels had to dump two-thirds of their cargo in order to navigate the same route. On average, one collier estimated, coal barges on the James ran at less than 70 percent capacity. When the Commonwealth of Virginia purchased a controlling interest in the James River Company in 1820, the legislature authorized the improvement of the route to allow the passage of large coal barges. Despite constant petitioning and pleading from coal interests, the James River Company never completed this

27. J. P. Pleasants to Harry Heth, July 8, 1811; Thomas B. Main to Harry Heth, December 2, 1815, Heth Papers, UVA; Chesterfield County Petition, December 15, 1824, Legislative Petitions File, Library of Virginia, Richmond, Virginia, hereinafter LVA.

work. In 1834, a group of colliers complained that the company neglected their interests and that "their businesses in the future must be carried on under great disadvantages, if not wholly abandoned." Appeals to build an entirely new coal-carrying route along the south bank of the James or to force local turnpikes to reduce tolls on coal fell upon deaf ears. In fact, in 1832 Virginia's legislature acquiesced to the Manchester Turnpike's demand to double its rates on coal wagons. Without the political clout of the state's agricultural interests, colliers found themselves marginalized in Richmond's transportation network, and the James River never developed an effective system for transporting coal from the mines of the Richmond Basin to the deepwater port at Rockets. Traffic down the James peaked in 1849 when the company sent nearly 38,000 tons of coal to tidewater, but declined soon thereafter, and by the eve of the Civil War, producers shipped only a little over 25,000 tons annually.[28]

Railways offered a potential alternative to the James River and its many shortcomings, especially for miners on the south side of the James. As with water transportation, Richmond Basin colliers could draw upon the British coal trade for inspiration. As early as the 1760s, collieries in the Newcastle and Sheffield areas used iron railroads to transport coal from the mouth of a mine to a staith. These simple lines, known as "self-acting railways," used gravity as a means of locomotion; as one full car traveled down the hill, an empty one attached by a pulley system climbed back to the mine. As an alternative to a purely gravity-based system, some British colliers used horses to pull one load as the empty container returned to the mouth of the mine.[29]

Ideas for constructing a railroad to service Chesterfield County coal mines circulated throughout the antebellum period. In 1812, John Stevens of New York wrote to Harry Heth and proposed that he invest in a thirteen-mile railroad along the south side of the James River.[30] Heth never acted on this plan, and frustration among

28. Chesterfield County Petitions, December 23, 1818, December 15, 1824, January 31, 1828; January 17, 1834, Legislative Petitions File, LVA; Wayland Fuller Dunaway, *History of the James River and Kanawha Company* (New York: Columbia University Press, 1922): 66-68; *Laws of Virginia, 1833* (Richmond: Thomas Ritchie, 1833), 95; "James River (Va.) Coal Trade," *Hunt's Merchant Magazine* 43 (December 1860): 753.

29. See, for example, Holland, *The Collieries and Coal Trade of Great Britain*, 353–58.

30. Nothing materialized from this plan, but it does suggest that alternative

coal interests blossomed in subsequent years. "The colliers of Chesterfield despair of being able to sustain the competition, unless they can very [naturally] reduce the price of transportation, from the pits to tide water," one petition argued in 1827, "and they perceive no other means of doing this, than the contemplated rail road."[31] The first serious consideration of a rail link from coal mines to market arrived in the form of Claudius Crozet's 1828 work entitled *Report on the Rail-Way from the Coal-Pits to James River.* In response to agitation for a railroad among the Chesterfield County coal interests during the 1820s the legislature instructed Crozet to investigate the possibility of constructing a rail link from the Chesterfield coal mines to the James River. Crozet concluded that a state-aided railway would reduce the cost of transporting Chesterfield coal from 9 cents to 3.5 cents a bushel, which would "actually save the coal trade, which otherwise will soon be unable to compete with that of the northern States." Crozet suggested that, rather than constructing a line from the coal pits to Manchester directly opposite the river from Richmond, a line connecting the mines to the James River would be the most cost-effective way to transport coal. Crozet's plan to construct a railroad for $55,450 also included a $90,000 improvement to the locks and loading equipment of the James River Canal.[32] This plan, while palatable to Chesterfield colliers, was improbable, given the James River Company's emphasis upon the upcountry agricultural trade and the reluctance of the state to invest more time and money in a system that was already far behind schedule. A state-funded railroad was therefore out of the question, and once again the interests of the plantation trumped the interests of the mine.[33]

Crozet's report did, however, result in the eventual construction of some railways that served the coal trade of the Richmond Basin.

---

methods of transporting coal were in the works at a very early time. John Stevens to Harry Heth, December 31, 1812, Harry Heth Papers, UVA.

31. Chesterfield County Petition, December 10, 1827, Legislative Petitions File, LVA. See also petitions from Chesterfield County dated December 22, 1825, Legislative Petitions File, LVA.

32. Claudius Crozet, Report on the Rail-Way from the Coal-Pits to the James River (1828): 4, 5; appended to the *Virginia Journal of the House of Delegates, 1827– 1828* (Richmond: Thomas Ritchie, 1828).

33. The sluggish progress of the James River improvements under state control spurred the transfer of the entire system to the James River and Kanawha Company, a state-supported joint-stock enterprise organized in 1835. Dunaway, *James River and Kanawha Company,* 90–91.

By 1831 the Chesterfield Railroad Company linked Manchester on the south side of the James River to the Midlothian Coal Pits, a distance of more than fifteen miles, but this line was abandoned in 1851. Other rail lines were small-scale efforts linking individual mines to larger improvements. For example, the Tuckahoe and James River Railroad Company built a five-mile link between the Tuckahoe Creek Mines and the James River in 1840. Some railroads that specialized in this kind of coal traffic appeared but failed to make a real impact on the region's transportation network. The Chesterfield and James River Railroad Company (1836), the Etana Coal Company (1836), and the Powhatan Railroad Company (1839) all received charters, but ultimately they constructed less than six miles of railways in the Richmond Basin. Other than the Chesterfield Railroad, it appears that rail lines in the Richmond Basin failed to reduce the cost and inconvenience of transporting coal to market. By the 1840s, the Clover Hill and the Richmond and Danville railroads shipped significant amounts of bituminous coal to local customers such as the Tredegar Iron Works, but by this time, Richmond Basin coal had been squeezed out of most coastwise markets. Railroad transport appeared too late to help establish Richmond Basin coal as a national commodity.[34]

If savings on transportation could not lower costs enough to make Virginia coal competitive in urban markets, perhaps limiting supplies would work. Once again, the British coal trade offered one potential model for America's nascent industry. During the eighteenth century, many mine owners in England's northeastern coalfields were organized in the Limitation of the Vend, a cartel designed to limit shipments of coal. This strategy raised coal prices in London by about 10 percent, even though members of the limitation could not completely control the entry of new colliers in the Newcastle region. By the time the cartel disintegrated in the early 1800s, the limitation had facilitated a dramatic reorganization of the Newcastle coal regions, as artificially high prices afforded increased investments in new pumping and raising methods. By digging deeper shafts and exploiting more extensive holdings, Newcastle colliers entered the nineteenth century poised to capture economies

34. "James River (Va.) Coal Trade," 753; Bruce, *Virginia Iron Manufacture*, 263-64; see also appendix in Christopher T. Baer, *Canals and Railroads of the Mid-Atlantic States, 1800–1860* (Wilmington, DE: Regional Economic History Research Center, 1981).

of scale and thus earn large profits without the protection of the Limitation of the Vend.[35]

Although such cartels would be difficult to construct in the United States, corporate charters offered one way to accumulate holdings and control prices in the American context. In this regard, the Heth family again led the way in the Richmond Basin. Harry Heth's son John and his partners successfully petitioned the legislature for a charter in 1833 to create the Black Heath Company of Colliers, which was the first coal-mining corporation in Virginia. But Heth's leadership here was not received warmly. Some smaller colliers complained that Heth's 1833 charter demonstrated "no privilege can be *given* to one or more Citizens that does not to the same degree *take* from the rights & privileges of their fellow Citizens" and worried that Heth's actions would open their trade to chartered companies, "the members of which may reside out of the state, and who by their vested rights and personal irresponsibility cannot be liable to the ordinary Legislation of the State." A few weeks later, as Heth's charter worked its way though the Virginia legislature, the opponents of the act petitioned again. "No matter how guarded the Language by which such powers are conveyed, or the purity of intention with which they are asked or granted," they warned, "we have abundant evidence that they are frequently perverted, and made subservient to purposes not contemplated by those who were instrumental in their Creation." The miners also noted that only one corporation mined anthracite in Pennsylvania and that for several years the legislature "has been taken up in curtailing its powers, and counteracting its influence; and have therefore repeatedly refused to charter any other company."[36]

The limited provisions of corporate charters in the Richmond Basin coal trade undercut the ability of large colliers to coordinate

---

35. For more on the Limitation of the Vend, see Peter Cromar, "The Coal Industry on Tyneside, 1771–1800: Oligopoly and Spatial Change," *Economic Geography* 53 (January 1977): 79–94; and William J. Hausman, "Cheap Coals or Limitation of the Vend? The London Coal Trade, 1770–1845," *Journal of Economic History* 44 (1984): 321–28.

36. *Laws of Virginia, 1832–1833*, 133–36; Eugene M. Scheel, *Culpepper: A Virginia County's History through 1920* (Culpepper, VA: Culpepper Historical Society, 1982), 126–27; Legislative Petition from Henrico County, December 11, 1832, Legislative Petitions Collection, LVA. Petitioners from Goochland County also invoked the idea that coal mining had been successfully accomplished in Great Britain by individual concerns for years and used the Pennsylvania Schuylkill anthracite region's rapid growth as an example of the productivity of individual entrepreneurs. Legislative Petition from Goochland County, January 5, 1833, Legislative Petitions File, LVA.

production and manipulate prices in urban markets. For example, the Black Heath Company of Colliers' charter made no provisions for expanding the land holdings of the firm and only incorporated the property already held by John Heth and his partners, Beverley Randolph and Beverley Heth. Once again, the Heth family set the pace for the Richmond Basin, but it became increasingly apparent that the charters used by Richmond colliers were not designed to restructure their trade. In 1834 Abraham Wooldridge applied for a charter so that he could form a mining firm that could survive the death of one or more partners. The charter of the Cold Brook Company of Colliers (1835) contained a restriction on purchasing additional lands and divided the initial shares among the women and children of the Cunliffe family. In that case, the corporate charter served as a way to consolidate a number of coal-mining tracts tied to dower or inheritance rights. Elizabeth Branch formed two corporations, the Dutoy and Powhatan Coal companies, with her lessees on two separate tracts of coal land, thus uniting both parties of the lease under a single corporate concern. These kinds of charters reinforced the status quo in the Richmond Basin and never facilitated any significant reorganization of the coalfield.[37]

The limitations on mining charters in the Richmond Basin restricted the economic growth of the region in many ways. Without the ability to acquire more land and raise capital, incorporated colliers could not sink investments into either northern or British technology. The corporate ownership of slaves might have mitigated some of the chronic labor shortages of the region, but the restrictions on raising capital found in early Virginia coal company charters made this strategy impossible. Even if colliers replicated the Limitation of the Vend in American fuel markets, it is unlikely that the rise in price would have kept Richmond Basin coal selling in urban markets. In fact, such action might have made Virginia coal susceptible to foreign competition. Tariff levels, which stood at six cents per bushel from 1825 to 1842 (about one-fifth of the price

---

37. Heth's firm was, however, authorized to purchase up to one thousand acres of land for timber. See *Laws of Virginia, 1832–1833*, 134. Wooldridge's firm, the Midlothian Coal Company, received a charter in 1835. See Legislative Petition from Chesterfield County, December 19, 1834, Legislative Petitions Collection, LVA. The Cunliffes' company was authorized to sell a small portion of shares at the company's formation, but the vast majority of interest in the company remained within the Cunliffe family. See *Laws of Virginia, 1834–1835*, 175–79; *Laws of Virginia, 1839–1840*, 119; *Laws of Virginia, 1840–1841*, 151.

of a bushel of Richmond bituminous), remained important to Richmond colliers. Should prices of domestic coal rise, British imports might once again claim a significant share of fuel markets in seaboard cities. Richmond's coal trade relied upon price competition aided by federal tariffs. "[T]he competition to which the Virginia colliers will be exposed if the duties are removed," a group of Richmonders petitioned Congress in 1837, "must of necessity be ruinous in the present state of affairs."[38]

The national significance of Virginia's early coal trade decreased precipitously with the rise of Pennsylvania anthracite in the 1820s and 1830s. Although the technical challenge of burning "hard coal" presented problems to anthracite colliers before the War of 1812, a concerted effort on the part of local entrepreneurs, Philadelphia's technological community, and Pennsylvania's state government had allowed anthracite to make significant gains in urban markets by the 1820s. As anthracite's brighter and hotter flame won over consumers, the future of bituminous coal waned. Annual production in the Richmond area surpassed 200,000 tons in 1835, but declined during the 1840s and 1850s to hover between 100,000 and 140,000 tons annually. Anthracite's rise to prominence, moreover, was wedded to the effective use of canal transportation, steam engines, and corporate charters. One anthracite canal, the Schuylkill Navigation Company, shipped over half a million tons of coal in 1837. By 1850, a single county in Pennsylvania's anthracite fields contained 169 engines of nearly 5,000 horsepower, a number that nearly doubled by 1855. In 1865, that same county used nearly 800 steam engines, which exceeded the power of all the stationary steam power in the United States reported in 1838 and contained fifty-two corporate mining firms using their charters as a way to raise capital and implement new technology in their mines. As the success of Pennsylvania anthracite demonstrated, Heth and other Virginia colliers had the right blueprint, if not the best environment, for success in the coal trade.[39]

What does the early history of America's first coal trade tell historians? In the short run, the story of the Richmond Coal Basin

38. "Coal Trade—Richmond," U.S. House Document No. 93, *24th Cong., 2d sess, 1836–1837* (Washington: Government Printing Office, 1837), 6.

39. Louis Hunter and Lynwood Bryant, *A History of Industrial Power in the United States, 1780–1930*, vol. 3, *The Transmission of Power* (Cambridge: MIT Press, 1991), 418–19; Eavenson, *American Coal Industry*, 440–44. For more on the early rise of

suggests that although the diffusion of knowledge of industrial practices might cross easily over national boundaries, the replication of a particular trade or industry as a complete system is a difficult endeavor. Coal companies duplicating the success of British mining methods eventually fell victim to the vagaries of plantation slavery. While the diffusion of technology occurred quite quickly in the abstract, social and political contexts determined success in practice. In his study of technological transfers from Europe to North America, Darwin Stapleton suggested that although innovations in internal improvements, gunpowder manufacturing, and iron making may have traveled successfully across the Atlantic, "transfers are not 'once-and-for-all' events but successive instances" that require "considerable social support." Virginia's slave-labor society, as Heth and his colleagues in the Richmond Basin discovered, offered a poor framework for such a prolonged technology transfer.[40]

The lack of coordination in the Richmond area and the failure of the James River Company, in both its public and private forms, to address the needs of the coal industry also made Virginia bituminous more expensive and less competitive in eastern markets. The spiderweb of turnpikes, canals, river improvements, and railroads could not significantly reduce the cost of transportation for local colliers. The political power and prestige of the James River and Kanawha Company, moreover, made other large-scale projects unfeasible, exacerbating the high cost of Virginia bituminous. Perhaps a window of opportunity existed during the 1820s and 1830s when Virginia coal was still an acceptable fuel in urban markets. It was during these decades, however, that the improvements and toll structure along the James River most favored tobacco and other agricultural products. Indirectly, the presence of slave-raised crops

anthracite coal in urban markets, see Frederick Binder, *Coal Age Empire: Pennsylvania Coal and Its Utilization to 1860* (Harrisburg: Pennsylvania Historical and Museum Commission, 1974); and H. Benjamin Powell, *Philadelphia's First Fuel Crisis: Jacob Cist and the Developing Market for Pennsylvania Anthracite* (University Park: Pennsylvania State University Press, 1978); Sean Patrick Adams, *Old Dominion, Industrial Commonwealth: Coal, Politics, and Economy in Antebellum America* (Baltimore: Johns Hopkins University Press, 2004), 48–83.

40. Darwin Stapleton, *The Transfer of Early Industrial Technologies to America: Memoirs of the American Philosophical Society Held at Philadelphia for Promoting Useful Knowledge*, vol. 177 (Philadelphia: American Philosophical Society, 1987), 28, 29.

influenced the development of effective transportation in the region. The Richmond Basin's advantageous position was ruined, and it failed to expand its influence and profitability. *"And will Virginia still sleep?"* cried the *Richmond Enquirer,* "While other states are pushing on with such gigantic strides, why do we loiter in the path?"[41]

In the long run, skeptics might argue, the Richmond Basin was limited by geographic size and undoubtedly would have been over-shadowed by colliers in the large anthracite and bituminous fields to the north and the west. Or perhaps the region's geology was sim-ply not conducive to extensive mining. Later reports suggest that this was not the case and instead that its early history, not necessar-ily its size or geological standing, undermined the region. When Samuel Harries Daddow and Benjamin Bannan published a com-pendium of American mineral resources, *Coal, Iron, and Oil; or the Practical American Miner,* in 1866, they included a dismal assessment of the Richmond Basin's history. "Experience and capital would undoubtedly remove some of the expense, and render mining more profitable," they wrote, "but the formations of this coal-field are so peculiar and uncertain, that no man, however experienced in other coal-fields, should feel confident in this, without much study and investigation." The area's shifting seams and slate-laden coal pre-sented problems, without a doubt, but miners continued to raise coal in the Richmond Basin throughout the nineteenth century. Nature could not be the only culprit.[42]

Nineteenth-century observers always mentioned the rude state of technology and the poor organization of the operations in the Richmond Basin, which suggests that the potential for expansion always existed. Daddow and Bannan saved the harshest invective for the Richmond Basin colliers themselves, by claiming that they had "gone backwards in the last ten years" and "insist on the bucket being the best and cheapest mode of drainage, and keep *on raising water instead of coal. Cui bono?"* Oswald Heinrich, the superintending min-ing engineer of the Midlothian colliery in the 1870s, argued that the region's problems stemmed from "ignorance, want of system, and false economy," among eastern Virginia mining firms, which

41. Reprinted in *Niles' Register,* February 21, 1829.
42. Samuel Harries Daddow and Benjamin Bannan, *Coal, Iron, and Oil; or, the Practical American Miner. A Plain and Popular Work on Our Mines and Mineral Resources, and a Text-Book or Guide to Their Economic Development* (Pottsville, PA: Benjamin Bannan, 1866), 397

"concentrated the worst elements imaginable to prevent the continuous prosperity of the mines." "An unaccountable lethargy seems to have fallen on all who touched this field," the *Engineering and Mining Journal* wrote of the Richmond Basin in 1876, "and yet, in the future it can scarcely fail to become one of the great sources of coal supply."[43]

So instead of a region doomed from the beginning, the wasted potential of the Richmond coalfield stands as an important "road not taken" in the economic development of Virginia, as a more dynamic coal industry might have spurred both forward and backward linkages in terms of technological advances, industrial development, and economic growth. Past and present scholars attribute this sectional distinctiveness of Virginia's agrarian economy to various factors, including entrepreneurial culture, urban development, and even differences in soil and climate.[44] Technology transfers that aided the growth of slavery-based crop cultures such as that surrounding rice, for example, found purchase in the institutional context of the American South. This suggests that slave labor itself was not the problem in transplanting coal mining to the Richmond Basin. But the struggle of Virginia's colliers to replicate Britain's dynamic coal industry illustrates the difficulties of engaging in industrial pursuits amidst plantation slavery. Harry Heth's large landholdings and social connections eventually helped propel his family into the upper reaches of the Old Dominion's gentry; his ties to Old World mining methods could not, however, push the Richmond Basin coal trade to the high ranks of American industry.[45]

43. Ibid., 402; Oswald J. Heinrich, "The Midlothian Colliery, Virginia," *Transactions of the American Institute of Mining Engineers* 1 (May 1871–February 1873): 348, and see also 360-64; and *Transactions of the American Institute of Mining Engineers* 4 (May 1875–February 1876): 308–16; *Engineering and Mining Journal* (January 1, 1876).

44. See, for example, Thomas Doerflinger, *A Vigorous Spirit of Enterprise: Merchants and Economic Development in Revolutionary Philadelphia* (New York: W. W. Norton, 1986), 335-64; Frederick Siegel, *The Roots of Southern Distinctiveness: Tobacco and Society in Danville, Virginia, 1780–1865* (Chapel Hill: University of North Carolina Press, 1987); Majewski, *A House Dividing*, 2000.

45. On the successful transfer of rice cultivation techniques, see Joyce Chaplin, "Tidal Rice Cultivation and the Problem of Slavery in South Carolina and Georgia, 1760–1815," *William and Mary Quarterly* 49 (1992): 29-61; Judith Carney, "Landscapes of Technology Transfer: Rice Cultivation and African Continuities," *Technology and Culture* 37 (1996): 5–35; Peter A. Coclanis, "How the Low Country Was Taken to Task: Slave-Labor Organization in Coastal South Carolina and Georgia," in David L. Carlton and Peter A. Coclanis, *The South, the Nation, and the World: Perspectives on Southern Economic Development* (Charlottesville: University of Virginia Press, 2003), 24–48.

Slavery and Technology in Louisiana's Sugar Bowl

RICHARD FOLLETT

A SUGAR PLANTATION, WROTE JEAN JACQUES AMPÈRE, WAS BOTH an agricultural and manufacturing concern; as the Frenchman saw it, Louisiana's sugar industry demonstrated that enslaved labor was indispensable for the region's cane lords and that sugar, more than any other crop, required the maintenance of racial bondage. Returning to his post at the College de France, Ampère grimly concluded that in the central question of mid-nineteenth-century humanity—namely the maintenance or abolition of slavery—his sojourn in south Louisiana's cane fields provided compelling evidence for those who wished to maintain the South's peculiar institution. Visitors from Europe and the northern states concurred, noting that without slavery, "we should have to resign altogether the production of sugar and rice, until we have reared in starving poverty a Paria class of whites miserable enough to undertake it." Most planters agreed and viewed sugar work firmly through the prism of racial slavery. Indeed, as the Englishman William Howard Russell candidly observed, "nothing but 'involuntary servitude' could go through the toil and suffering required to produce sugar." On the estates he visited, Russell encountered a merciless industry in which cane farming and mechanized sugar production fused on agroindustrial plantations. The slaves—he noted—conducted "all the work of skilled labourers," and the regimented field gangs

The author wishes to express his gratitude to the Arts and Humanities Research Board of the United Kingdom for funding the project "Race and Labour in the Cane Fields: Documenting Louisiana Sugar, 1844–1917" (APN No: 16,426) from which this paper derives.

moved through the sprouting canes with almost military discipline. Like Ampère, Russell was impressed by the modernity of the estates though wholly shocked by the total and structured exploitation of African American laborers in Louisiana's sugar country.[1]

For centuries, the meandering bayous of the Mississippi floodplain had been home to native Americans who had adapted their agriculture to the vagaries of their alluvial landscape and the perennial risk of spring freshets and inundation. The rivers and bayous that snaked through the region, however, constantly replenished the rich soils, leaving narrow corridors of highly fertile land that followed the crescents and bends of Louisiana's dense network of watercourses. Despite its advantages, the humid climate and lush soils of southern portions of the state were ill-suited for short-staple cotton, the antebellum South's principal crop. The fertile alluvium, nonetheless, proffered wealth for those who could tame the habitat and convert "waste lands into verdant fields," as one nineteenth-century writer observed, and reap "stores of gold and silver from the glebe they turned up." And harvest its riches they surely did, for in the sixty years following Etienne Boré's successful granulation of Louisiana sugar in 1793, planters and slaves converted the river crests and levee fronts into a network of plantations that stretched from the Gulf of Mexico in the South to Rapides Parish in central Louisiana and to St. Mary Parish in the West.[2]

Initially growing cane in the New Orleans hinterland, planters swiftly expanded along the Mississippi River to Baton Rouge. With hardier cane varieties increasingly available and steam-powered mills replacing those powered by animals and reducing the length of time it took to grind canes, sugar planters successfully cultivated lands in the hill country north of the state capitol and farmed cane where it had previously been deemed unfeasible to do so. To the

---

1. Jean Jacques Ampère, *Promenade en Amérique* (Paris: M. Lévy frères, 1855), 137–38; Cora Montgomery, *The Queen of Islands and the King of Rivers* (New York: C. Wood, 1850), 35; William H. Russell, *My Diary North and South* (Boston: T.O.H.P Burnham, 1863), 259, 273.

2. *Franklin Planters' Banner*, March 16, 1848. On the historical geography of the cane world, see Sam B. Hilliard, "Site Characteristic and Spatial Stability of the Louisiana Sugarcane Industry," *Agricultural History* 53 (January 1979): 254-69; John B. Rehder, *Delta Sugar: Louisiana's Vanishing Plantation Landscape* (Baltimore: Johns Hopkins University Press, 1999), 1–59. On early Louisiana, also see Daniel H. Usner Jr., *Indians, Settlers, and Slaves in a Frontier Exchange Economy: The Lower Mississippi Valley before 1783* (Chapel Hill: University of North Carolina Press, 1992), 149–90.

west, the antebellum sugar industry similarly expanded along Bayous Teche and Lafourche and the many tributaries that fed into them. Reasonable cane prices combined with federal tariff protection encouraged further experimentation in the 1840s as the industry gradually expanded, occupying new lands, draining swamps, and reclaiming the marshy landscape for the sugar masters. By the eve of the Civil War, the antebellum industry had reached its geographic and production limits. Ultimately constrained by the volume of available land and by cane sugar's susceptibility to frost, the industry nonetheless posted remarkable growth. By midcentury, some 125,000 enslaved African Americans toiled on 1,500 estates producing 250,000 hogsheads of raw sugar. In the following decade, planters consolidated their holdings, smaller producers left the industry, and the total number of sugar estates declined to 1,300. Consolidation, however, did not check overall expansion. In fact, by 1853, Louisiana planters were producing a quarter of the world's exportable sugar. Enthusiastically fanning regional pride, Representative Miles Taylor announced that such progress "is without parallel in the United States, or indeed in the world in any branch of industry." After several poor crops in the mid-1850s, production recovered in the final antebellum years, and as the nation descended into Civil War, planters celebrated the bumper 1861 harvest of 460,000 hogsheads. It was the last crop made entirely by slave labor and the crowning moment of the antebellum sugar masters.[3]

3. *De Bow's Review* 1 (January 1846): 55–56. On the growth of the sugar industry, see Richard Follett, *The Sugar Masters: Planters and Slaves in Louisiana's Cane World, 1820–1860* (Baton Rouge: Louisiana State University Press, 2005); Richard Follett, "On the Edge of Modernity: Louisiana's Landed Elites in the Nineteenth-Century Sugar Country," in *The American South and the Italian Mezzogiorno: Essays in Comparative History,* ed. Enrico Dal Lago and Rick Halpern (Basingstoke: Palgrave, 2002), 73–94; J. Caryle Sitterson, *Sugar Country: The Cane Sugar Industry in the South, 1753–1950* (Lexington: University of Kentucky Press, 1953), 13–44; P. A. Champomier, *Statement of the Sugar Crop Made in Louisiana, 1845–1846* (New Orleans: Cook, Young, & Co., 1846), 35; *Statement of the Sugar Crop, 1849–1850,* 51; *Statement of the Sugar Crop, 1859–1860,* 39; *Statement of the Sugar Crop, 1861–1862,* 39. On northern expansion and increased yields, see *De Bow's Review* 2 (December 1846): 442; *De Bow's Review* 3 (May 1847): 414; *Alexandria Democrat* reprinted in *Planters' Banner,* January 13, 1848; *American Agriculturist* 9 (November 1850): 351. Postwar events are summarized in Richard Follett and Rick Halpern, "From Slavery to Freedom in Louisiana's Sugar Country: Changing Labor Systems and Workers' Power, 1861–1913," in *Sugar, Slavery, and Society: Perspectives on the Caribbean, India, the Mascarenes, and the United States,* ed. Bernard Moitt (Gainesville: University of Florida Press, 2004), 135–56; Louis Ferleger, "Farm Mechanization in the Southern Sugar Sector after the Civil War," *Louisiana History* 23 (winter 1982): 21–34; but

But until the introduction of modern frost-resistant canes, cane farming on the lower reaches of the Mississippi proved a perilous and risky pursuit. As William P. Bradburn of the *Plaquemine Southern Sentinel* underscored, the risk of misfortune ultimately defined antebellum sugar farming. "In our countryside," he observed, "the people seem run mad upon the culture of staple products. . . . They turn the farmers' life into that of a gambler and speculator. They are dependent upon chance, and an evil turn of the cards—a bad season, a fall in prices, or some such usual calamity." Despite such grave counsel, Bradburn observed, planters continued to "run mad" funneling vast sums into land, slaves, and sugar production equipment. Located on the northern rim of the Caribbean sugar-producing belt, Louisiana faced a series of overlapping meteorological problems. Icy winds from Minnesota swept down the central corridor of the nation, bringing sharp frosts that froze the sucrose-rich juice within the canes and rendered it all but worthless. Cane growers faced an almost impossible dilemma: if they planted the majority of their seed crop in January and harvested it some nine or ten months later, the cane was immature, with lower sucrose content than that enjoyed by their Caribbean rivals; if they waited a few weeks more, the canes matured, but the planters risked their fortunes against the climes. Most planters ultimately chose to harvest in mid- to late October. After an eight-month growing season and facing a volatile climate, slaves worked relentlessly to gather and process the canes before the first killing frosts descended. Once the grinding season began, operations continued round the clock as cane cutters advanced over the fields supplying the mill with freshly cut canes at breakneck speed. Slaves then fed the canes through the mill and extracted the sugar juice. The evaporation process took place over four open kettles, which varied in size; at the smallest, or battery, granulation began. As a roaring furnace kept the kettles at the correct temperature, skilled sugar makers added lime to the boiling juice to remove impurities. Slaves then skimmed the clarified juice and ladled it into the next kettle for evaporation. In the last kettle, the sugar maker carefully watched the cane juice begin to granulate, its thick grainy appearance

---

above all, see John Rodrigue, *Reconstruction in the Cane Fields: From Slavery to Free Labor in Louisiana's Sugar Parishes, 1862–1880* (Baton Rouge: Louisiana State University Press, 2001).

signaling the final and crucial stage of production. The sugar maker would then order the semimolten juice to be struck. Slaves transferred the clarified sugar into wooden vats, known as coolers, where the sugar would crystallize over the course of twenty-four hours. It was then packed into large wooden hogsheads (containing between one thousand and twelve hundred pounds of sugar) with holes in their bases and left to drain until the molasses separated from the brown crystalline sugar. Planters then either shipped the sealed hogsheads to market to be consumed raw or to one of several northern cities to be refined. At every point in the harvest production schedule, planters held speed at an absolute premium as they strove to fashion reliable work crews who would efficiently harvest the annual crop and synchronize field operations with those in the mill house.[4]

While the gruelling pace of the agricultural year peaked during the harvest months, plantation work continued unchecked during the spring and summer. Immediately after the cane was processed in late December, slaves began seeding and planting the next year's crop—a task that seldom matched the intensity of harvest labor but one that nonetheless involved extensive and arduous work. Drainage canals, moreover, required constant maintenance before plowhands furrowed the soil into perpendicular rows, which were planted with seed cane. Until mechanization in the twentieth century, planting cane required backbreaking labor through the late winter months as farmers hurried to prepare their cultivation ridges and thus maximize the already foreshortened growing season. Not every field required replanting. Like many members of the grass family, sugarcane is a perennial and in warm tropical climes yields up to six crops. In cooler Louisiana, however, planters allowed the cane to ratoon (resprout from the plant base) for only two years before replanting. To be sure, first- and second-year ratoons required less attention in the postharvest months, but throughout the region, the sapping pace of agricultural life did not relent as planters drove the slaves to sow the canes as swiftly as

---

4. *Plaquemine Southern Sentinel,* June 22, 1850. On cultivation, harvesting, and processing, see Walter Prichard, "Routine on a Louisiana Sugar Plantation under the Slavery Regime," *Mississippi Valley Historical Review* 14 (September 1927): 168–78; Glen R. Conrad and Ray F. Lucas, *White Gold: A Brief History of the Louisiana Sugar Industry, 1795–1995* (Lafayette: University of Southwestern Louisiana Press, 1995), 14–22.

possible. In the spring and summer, slaves continued to toil in the fields, chopping away the weeds that grew among the sprouting canes. By mid-June or early July, the shoots were robust enough to survive without constant attention, though even during this "lay-by" season, slaves turned to maintenance tasks, namely the cultivation of corn, levee repairs, and the collection of timber to fuel the steam-powered sugar mill come harvest. The overlapping duties placed continuous pressure on the slave crews, who worked to a strict production agenda where time constraints dictated both the grinding and planting seasons. Speed and workplace reliability accordingly commanded immediate attention as few estate managers could afford the potentially disastrous impact of delays in a very time-conscious industry.

To address these related concerns, Louisiana's antebellum elite adopted a rigorous and disciplined labor regime and introduced labor- and time-saving machinery on their estates. In so doing, they established the American South's most industrialized "agricultural business" and one that only the richest of Caribbean sugar lords could match in terms of capital investment. But the path toward technical modernity was strewn with obstacles—the unpredictability of harvest yields, the mounting costs, growing international competition, and above all, a guarded wariness toward the enormous expenses of industrial sugar production. Such cautiousness on the behalf of the planter class was not unfounded. Until the introduction of centralized grinding facilities in the 1880s, when cane farmers sold their cane to a collective mill for rolling, almost all Louisiana sugar planters possessed their own milling facilities. Since even modestly priced sugar mills proved relatively costly for all save the very rich, machinery investment proved to be an expensive, though essential outlay for every sugar producer. To be sure, some sugar planters quailed at the prospect of investing tens of thousands of dollars in the latest machinery, but the pressures to harvest and grind their crops at double-quick speed ensured that by midcentury most planters had grasped at the panacea offered by steam-powered milling. This was a major technical breakthrough, but the same planters who oversaw the steam revolution proved unwilling or slow converts to the hugely expensive vacuum-processing of sugar, a manufacturing technique that appeared in the 1850s but which was not fully adopted by Louisiana planters until the postbellum era. Although a number of elite sugar planters adopted vacuum processing and took the next technical step in the production of cane sugar, most cane farmers continued to produce sugar with the open-kettle technique but with the use of steam power for milling. Some planters additionally experimented with partnerships and mergers to enhance their capital bases, but even these lacked the financial liquidity to advance the industry substantially. Technically, therefore, the Louisiana sugar industry underwent two evolutionary stages during the slave era. The first featured the replacement of horsepower with steam power for the grinding of sugar and the introduction of a gang labor system that would provide the antebellum sugarhouse with a steady supply of canes. As this essay underlines, the introduction of steam power proved a major step in the evolution of the Louisiana sugar industry and of the lower Mississippi valley plantation belt. The second stage of technical evolution of the Louisiana sugar industry, however, gathered momentum slowly. Wealthy planters experimented with the vacuum evaporation of sugar in the 1850s, but it would be several decades more before cane industrialists moved ahead en masse with these innovations. The enormous costs incurred with vacuum processing and the volatility of the prewar industry certainly contributed to the slower pace of technical evolution in the 1850s, but the planters' desire for economic independence additionally ensured that Louisiana's cane elite funneled resources into private capital acquisitions, though less frequently cooperated with their neighbors to purchase expensive machinery as collectives or associations. While cost undoubtedly contributed to the planters' disinclination to purchase vacuum evaporators, it was also the singular dearth of associationalism in the sugar belt and the myopic individualism of the cane planters that underpinned the sporadic and rather sluggish approach many planters adopted to the next stage of economic evolution in the 1850s.[5]

5. Walter Brashear to Francis C. Brashear, September 16, 1833, Brashear and Lawrence Family Papers, Southern Historical Collection, Manuscripts Department, Wilson Library, University of North Carolina, Chapel Hill, hereinafter UNC. On antebellum technology and slavery, see Follett, *Sugar Masters*, 90–150; Sitterson, *Sugar Country*, 112–56; John Alfred Heitmann, *The Modernization of the Louisiana Sugar Industry, 1830–1910* (Baton Rouge: Louisiana State University Press, 1987), 8–48; Rodrigue, *Reconstruction in the Cane Fields*, 13–16. Postbellum developments are carefully addressed in Heitmann, *Modernization of the Louisiana Sugar Industry*, 49–114; Ferleger, "Farm Mechanization," 21–34.

Fortunately, the debate over whether the antebellum South was definitively capitalist or precapitalist no longer divides the historical community as it once did. Most scholars now accept that aspects of capitalism and premodern values existed in tandem. As I've argued elsewhere, nowhere was this truer than in the antebellum sugar country where Louisiana cane planters were doubtlessly capitalist in their economic vision and invested in highly developed plantations, but they simultaneously embraced a social ethic based on mastery, individualism, and independence. As a generation of scholarship has now shown, the culture of American slaveholding ultimately bred jealous independence, a myopic focus on the individual, one's own authority, and upon personal liberty. Acquisitive and market-oriented, expansionist slaveholders spoke a lingua franca of modernity but tempered it with paternalism, planter hegemony, and a regionwide commitment to the preservation of liberty, slavery, and republican precepts of independence and virtue. This social ethic placed a premium on individualism and personal mastery, be it over the land, slaves, or capital. Above all, the planters' self-identity remained anchored to the plantation and to their role as slaveholders and labor lords. Capital expenditure on a plantation mansion, more slaves, or the latest machinery heightened the planters' sense of mastery over land, labor, and sugar, though public cooperation did little to exalt their power. The sugar elite accordingly invested in more slaves or steam-powered mills, but they ultimately embraced a blinkered, estate-focused brand of southern capitalism. It enabled them to produce large crops and ruthlessly exploit their slaves, though as we shall see, it proved shortsighted in the long run. In particular, the planters' highly localized and rather insular vision of economic evolution provided little scope for the associationalism that recent historians have characterized as a leitmotif of capitalist societies. Rather, the planters' self-absorbed commercialism ensured that irrespective of individual plantation wealth and progress, the sugar parishes remained in a transitionary stage, pockmarked by hundreds of well-established, capital-intensive estates but ultimately lacking the infrastructure of a fully capitalist society. From the perspective of the Mississippi River sugarhouses, antebellum Louisiana appeared as a technically progressive island of the New South within the Old, but on further inspection, we find an industry and region, rooted not in collective progress, associationalism, or collaboration, but rather focused on individual self-sufficiency, independence, and private investment.[6]

Louisiana's sugar elite thus developed a commercial corollary to their broader plantation-based social ethic. It condoned independent private investment but shied away from associationalism and collaborative costs and equipment that might threaten their individual plantation-based sovereignty. Agricultural associations and state-wide institutions formed to disseminate technical expertise thus foundered on a bedrock of planter intransigence—certainly, the plantocracy read commercial magazines like *De Bow's Review* and ordered bigger and more powerful mills, but their vision of economic evolution rested wholly upon individual plantation success and the unitary producer utilizing his own harvesting facilities. Independence, of course, gave the planter a degree of leverage and autonomy in the market—he could decide when he thought best to harvest, he could personally oversee the making of sugar to produce the most valuable commodity for onward sale, he could dictate the pace of production and exercise his authority and sway over those who toiled in his pay

6. The principal works advocating an essentially noncapitalist perspective on planter identity include, but are not limited to, Eugene D. Genovese, *The Political Economy of Slavery: Studies in the Economy and Society of the Slave South* (New York: Random House, 1965); Eugene D. Genovese and Elizabeth Fox-Genovese, *Fruits of Merchant Capital: Slavery and Bourgeois Property in the Rise and Expansion of Capitalism* (New York: Oxford University Press, 1983); Eugene D. Genovese and Elizabeth Fox-Genovese, *The Slaveholder's Dilemma: Freedom and Progress in Southern Conservative Thought, 1820–1860* (Columbia: University of South Carolina Press, 1992). By contrast, Robert Fogel, Stanley Engerman, James Oakes, and William Dusinberre have stressed the profit orientation of slaveholders and the capitalist nature of slavery. In later works, however, Oakes, Dusinberre, and Shearer Davis Bowman argue that southern planters were intensely capitalist but not democratic, and they sought to maintain strict social hierarchies at the expense of liberal capitalism. See Robert W. Fogel and Stanley L. Engerman, *Time on the Cross: The Economics of American Negro Slavery* (New York: W. W. Norton, 1974); Robert W. Fogel, *Without Consent or Contract: The Rise and Fall of American Slavery* (New York: W. W. Norton, 1989); James Oakes, *The Ruling Race: A History of the American Slaveholders* (New York: Vintage Books, 1982); James Oakes, *Slavery and Freedom: An Interpretation of the Old South* (New York: Alfred A. Knopf, 1990); Shearer Davis Bowman, *Masters and Lords: Mid-Nineteenth-Century U.S. Planters and Prussian Junkers* (New York: Oxford University Press, 1993); William Dusinberre, *Them Dark Days: Slavery in the American Rice Swamps* (New York: Oxford University Press, 1996). On the southern fixation with liberty and independence, also see J. William Harris, *Plain Folk and Gentry in a Slave Society: White Liberty and Black Slavery in Augusta's Hinterlands* (Middletown, CT: Wesleyan University Press, 1985); Lacy K. Ford, *Origins of Southern Radicalism: The South Carolina Upcountry, 1800–1860* (New York: Oxford University Press, 1988). More recently, historians have presented southern slavery as a complex hybrid of capitalist and precapitalist values in which slave owners demonstrated capitalist pretensions with a labor system and social values that espoused paternalistic and

or under the threat of his lash. Autonomous production thus served multiple social, economic, and structural imperatives; it enhanced the master's sense of authority and command and offered a relatively practical, albeit rather costly, solution to the annual harvest rush. Local boosters, however, were not wholly convinced by the planters' individualistic production plans. As the Port Allen newspaper, *Capitolian Vis-à-Vis*, pointedly remarked: "there is neither union, co-operation or friendly association existing, each person pursuing his own plan, prosecuting his own theories, and perpetuating a great deal of mischief." Exasperatedly calling for collaboration and regional association, newspaper editors counseled the sugar masters to protect their interests from "ruinous competition" by prudent management and astute investment. Despite such counsel, the sugar masters retrenched in isolation, plowing their profits into stand-alone plantation enterprises and prosecuting their own plans for the production and sale of sugar.[7]

Irrespective of the judiciousness of independent action, commentators who traveled south to visit the sugar region frequently

patriarchal values; these include Christopher Morris, *Becoming Southern: The Evolution of a Way of Life, Warren Country and Vicksburg, Mississippi, 1770–1860* (New York: Oxford University Press, 1995); Mark M. Smith, *Mastered by the Clock: Time, Slavery, and Freedom in the American South* (Chapel Hill: University of North Carolina Press, 1997); Jeffrey R. Young, *Domesticating Slavery: The Master Class in Georgia and South Carolina, 1670–1837* (Chapel Hill: University of North Carolina Press, 1999); Daniel Dupre, "Ambivalent Capitalists on the Cotton Frontier: Settlement and Development in the Tennessee Valley of Alabama," *Journal of Southern History* 56 (May 1990): 215–40; James David Miller, *South by Southwest: Planter Emigration and Identity in the Slave South* (Charlottesville: University Press of Virginia, 2002); William Kauffman Scarborough, *Masters of the Big House: Elite Slaveholders of the Mid-Nineteenth-Century South* (Baton Rouge: Louisiana State University Press, 2003); Jonathan Daniel Wells, *The Origins of the Southern Middle Class, 1800–1861* (Chapel Hill: University of North Carolina Press, 2004); Follett, *Sugar Masters,* 151–94; Tom Downey, *Planting a Capitalist South: Masters, Merchants, and Manufacturers in the Southern Interior, 1790–1860* (Baton Rouge: Louisiana State University Press, 2006).

7. *Port Allen Capitolian Vis-à-Vis,* August 23, 1854. On the relative failure of associations, see Heitmann, *Modernization,* 25–48, Richard Follett, "'Give to the Labor of America, the Market of America': Marketing the Old South's Sugar Crop, 1800–1860," *Revista de Indias* 65, 233 (January–April 2005): 117–47. For interpretations that contrast capitalistic features in the South with nonmarket values, see for introductions to the extensive bibliography, Mark D. Smith, *Debating Slavery: Economy and Society in the Antebellum American South* (Cambridge: Cambridge University Press, 1998); Douglas R. Egerton, "Markets without a Market Revolution: Southern Planters and Capitalism," *Journal of Early Republic* 16 (summer 1996): 207–21; Harry L. Watson, "Slavery and Development in a Dual Economy: The South and the Market Economy," in *The Market Revolution in America: Social, Political, and Religious Expressions, 1800–1880,* ed. Melvyn Stokes and Stephen Conway (Charlottesville: University Press of Virginia, 1996), 43–73.

remarked on the agricultural and industrial transformation that the planters had orchestrated throughout south Louisiana. Charles Fleischmann concluded in his "Annual Report for the Commissioner of Patents" that "there is no sugar growing country, where all the modern improvements have been more fairly tested and adopted than in Louisiana." Attributing the success of these "improved modes" to the "enterprise and high intelligence of the Louisiana planters, who spare no expense to carry this important branch of agriculture and manufacture to its highest perfection," Fleischmann paralleled other observers in noting that, despite the climatic limitations to cane cultivation in south Louisiana, the planters achieved a "proud triumph" in adopting the latest boiling apparatus and in "fulfilling all the conditions that science and experience have pointed out . . . for obtaining a pure and perfect crystalline sugar." One anonymous contributor to the *Baton Rouge Gazette* similarly lauded his fellow sugar masters for their skilful mastery of "the mechanical and chemical sciences which now become so apparent in this country." Familiar with several "going-a-head" planters, the correspondent announced that by introducing improvements in agriculture and machinery, the sugar master "will reap his harvest in half the time, and with half the labor and expense" than he previously achieved with primitive agronomy and animal-powered sugar mills. One planter from Guadeloupe, after a tour through his native Caribbean, additionally observed that the Louisiana sugar country appeared "far superior to most sugar growing regions . . . in the intelligence and skill manifested in both the cultivation and manufacturing of sugar." Ever partisan, the *Planters' Banner* further added that the sugar masters displayed both "intelligence and skill" in their planting operations combined with "good management on the improved principle adopted in Louisiana." This blend of management and skill not only assured the relative economic success of the U.S. sugar industry, but, the St. Mary Parish paper concluded, it also gave Louisiana planters a marked advantage over their competitors in Mexico, Cuba, and the West Indies.[8]

8. Charles L. Fleischmann, "Report on Sugar Cane and Its Culture," U.S. Patent Office, *Annual Report of the Commissioner of Patents for the Year 1848. 30th Cong., 2d sess., House of Representatives Doc. No. 59* (Washington, DC: Wendell & Van Benthuysen, 1849), 275; *Baton Rouge Gazette*, December 2, 1843; *De Bow's Review* 15 (December 1853): 648; *Franklin Planters' Banner*, January 5, 1854.

Good land, impressive yields, and federal tariff protection ensured profit margins for the antebellum elite and sheltered the industry from the rigors of global competition. Although the sugar duty varied, these revenue-raising tariffs had the additional benefit of allowing Louisiana planters and merchants to increase the price of domestic sugar by at least the duty charged on imported goods. Thus, when Congress levied a duty of three cents per pound on foreign sugars in 1816, Louisiana planters gained as much as eight to ten cents per pound for their sugars. With the exception of the 1842 bill, subsequent tariffs reduced sugar protection to a 30 percent ad valorem rate. Predictably Whig in their politics, Louisiana planters badgered their representatives to maintain and augment the tariff, but for most of the antebellum years, the sugar duty guaranteed reasonably good profits of 6 to 12 percent and offset the notoriously heavy costs of cane sugar production in the state. High capital investment in land, labor, machinery, and supplies combined with comparatively poorer yields per acre than those of their competitors ensured that Louisiana sugar proved relatively expensive to produce. Since those production costs ranged from four to five and a half cents per pound and prices remained between six to eight cents a pound for most of the antebellum years, the federal tariff fundamentally cocooned Louisiana planters from ruinous competition and ensured relatively assured profits.[9]

Planters frequently complained about the insecurity of those margins, occasionally claiming that sugar farming "is essentially a gambling operation," but despite their protests, promising sugar returns ensured regional expansion as the sugar industry moved beyond its initial core in the southern and western parishes to incorporate formerly marginal cane land in the central portion of the state. Depressed cotton prices in the 1840s further stimulated planters to convert their operations to cane. Stretching north from Pointe Coupee Parish, cane cultivation thus spread as far as the Red River, where the growing season for sugarcane was at its absolute minimum and where autumn chills represented a permanent threat. Thus while speed was of the essence throughout Louisiana, rapid harvesting and grinding proved vital on the northern

9. Joseph G. Tregle Jr., "Louisiana and the Tariff, 1816–1846," *Louisiana Historical Quarterly* 25 (January 1942): 24–148, 30; Philip D. Shea, "The Spatial Impact of Governmental Decisions on the Production and Distribution of Sugar Cane, 1751–1972" (Ph.D. diss., Michigan State University, 1974), chap. 2.

margins of the industry. Despite their meteorological limitations, the *Alexandria Democrat* proudly trumpeted: "We have the soil, climate, wealth, and energy for the successful prosecution of this new branch of industry and the day is not far distant when Rapides will take rank at the head of the Sugar Parishes." Neighboring Avoyelles Parish similarly declared their superiority in cane cultivation, warning older sugar parishes to the south that "no portion of our great Republic is superior to Avoyelles." R. L. Allen reporting for *De Bow's Review* noted that "the extension of cane cultivation is undoubtedly advancing more rapidly at the present moment than at any former period. Each succeeding year witnesses the extension over new territory." So impressive was the increase in sugar cultivation to the north that the *American Agriculturist* announced:

> Baton Rouge, instead of being far above all the sugar plantations, is becoming a central point. The march of the cane has passed her many miles and leaving . . . the Mississippi, has taken position far back among the hills . . . Such has been the success of the last two years, that many new mills are being erected, and vast quantities of land brought into cultivation in places where it would have been thought madness to talk of making sugar ten years ago.[10]

Expansion on the Red River grew rapidly, but so too did sugar farming in the Felicianas and on reclaimed land at the rear of established plantations that could be drained for the future cultivation of cane. P. A. Roy, editor of the *Pointe Coupee Democrat*, aptly described the severity of the land pressure in 1860: "Lands suitable for the cultivation of . . . sugar, are rising so fast in value and the demand is so great," he added, "that it becomes absolutely necessary to bring into cultivation, by artificial means, those lands which have heretofore been looked upon as unfit for cultivation, but when brought in cultivation, are the richest in the State." Recovering cotton prices

10. Moses Liddell to John R. Liddell, July 28, 1845, Moses and St. John Richard Liddell Family Papers, Louisiana and Lower Mississippi Valley Collections, Hill Memorial Library, Louisiana State University (hereinafter LSU). On profits, see David O. Whitten, "Antebellum Sugar and Rice Plantations, Louisiana and South Carolina: A Profitability Study" (Ph.D. diss., Tulane University, 1970), 81–96; David O. Whitten, "Tariff and Profit in the Antebellum Sugar Industry," *Business History Review* 44 (summer 1970): 226–33; On growth of the sugar region, see *Alexandria Democrat* reprinted in *Franklin Planters' Banner*, January 13, 1848; *New Orleans National,* reprinted in *Franklin Planters' Banner*, September 30, 1847; *De Bow's Review* 3 (May 1847): 414; *American Agriculturist* 9 (November 1850): 351.

and the exorbitant costs of sugar drove a number of these northerly cane producers out of the industry in the early 1850s, but the growth of the late antebellum industry ensured that sugar farming extended across the state and provided a substantial market for agricultural technology.[11]

Although the pressing demand for swift harvesting technology compelled planters to innovate, the introduction of tougher cane varieties in the 1820s necessitated grinding mills that could exert greater pressure on the canes than previously had been possible and extract their sugar juice more efficiently. The tougher ribbon cane matured earlier than other varieties and proved more resistant to frost damage, but its bark was difficult to crush with animal-powered mills. The introduction of steam-powered milling in 1822 with its improved grinding capacity largely resolved this techno-logical difficulty and cracked the bottleneck to increased cultiva-tion. With milling facilities that could swiftly grind the crop, planters could extend their planting, confident that they possessed the technology to process the canes. During the 1820s, most planta-tions utilized mills that included two or three horizontally or verti-cally placed cylindrical rollers that turned under the power of several oxen. Although this technology had sufficed for earlier pro-ducers in the Caribbean, where the harvest was conducted over a three- to four-month period, it proved inferior in antebellum Louisiana with its relatively short grinding season. To crush the rib-bon canes and counteract the meteorological conditions that com-pelled Louisianans to roll their cane in half the time of their Caribbean neighbors, planters, in the first instance, improved their grinding machinery by adding a further roller and by reducing the gap between the cylinders. This modest and relatively inexpensive step enabled planters to extract greater juice and crush the canes, but extraction rates of 40 to 50 percent still disappointed. Moreover, the mills still relied on the plodding pace of a mule or an oxen to turn the mill shaft. To increase speed, planters could push on through the night and run their mills flat out or purchase more powerful mills that crushed the canes at an agroindustrial pace. Planters ultimately adopted all three of these strategies, though by the early 1830s, steam-powered mills increasingly replaced the slower, less efficient animal-powered mills. Steam power had many

11. *New Roads Pointe Coupee Democrat*, August 18, 1860.

advantages: it vastly improved the pressure mills could exert on the canes and added a mechanical tempo to work in the cane industry. This in turn commanded managerial changes as planters drove their slave crews to toil at the metered cadence of the industrial age. For those on the mill floor or in the cane fields, steam power translated into grueling workplace practices as planters converted their farms into agricultural factories that toiled round the clock.

Steam milling extended quickly, in no small measure due to the declining cost of grinding equipment and the rapid extension of ribbon cane into northern and western parts of the Louisiana sugar bowl. Introduced in 1822, steam-powered sugar mills proved initially very costly at approximately twelve thousand dollars, though by the 1830s, planters could reasonably expect to purchase a domestically built sugar mill for five to six thousand dollars. These prices fell still further with New Orleans–based Leeds and Company offering their cheaper mills at just over three thousand dollars by midcentury. In addition, a vibrant secondhand market assured that while still expensive, steam-powered grinding machinery remained within the grasp of moderately wealthy individuals. The gradual cheapening of steam power eased its swift integration within the regional sugar economy and by 1828, 120 of the 691 estates utilized steam engines to drive their mills. Thirteen years later, steam powered 361 of the 668 Louisiana sugar estates, and by 1850, steam engines were operating in over 900 plantations and grinding almost all commercially produced cane. By the end of the antebellum era, almost 80 percent of sugarhouses possessed steam engines and mills that could crush and grind ribbon cane with increasing efficiency. Those who retained animal-powered mills gradually found themselves edged out onto the geographic, technical, and economic margins of sugar production.[12]

Steam power, nonetheless, introduced new complications for the sugar plantocracy. First, in harmonizing the manual process of cutting canes with the mechanized production of sugar, nineteenth-

12. On the transformation from animal to steam power, see *De Bow's Review* 1 (January 1846): 55; Fleischmann, "Report," 294; *Franklin Planters' Banner*, July 29, 1847; Edward J. Forstall, *Agricultural Productions of Louisiana, Embracing Valuable Information Relative to the Cotton, Sugar and Molasses Interests, And the Effects upon the Same of the Tariff of 1842* (New Orleans: Tropic Print, 1845), 4; J. A. Leon, *On Sugar Cultivation in Louisiana, Cuba, etc., and the British Possessions by a European and Colonial Sugar Manufacturer* (London: J. Ollivier, 1848), 26; Champomier, *Statement of Sugar Crop in 1850–1851*, 43; *Statement of Sugar Crop in 1860–1861*, 39.

century sugar growers faced a challenging disequilibrium between matching the pace of the cane cutters to the mechanical pace of the sugar mill. To do this, Louisiana's sugar planters combined elements of modern industrial development with the traditions of sugar cultivation. To visitors like Ampère, mid-nineteenth-century sugar plantations seemed almost military in their industrial organization. Indeed, James Ramsay's eighteenth-century conviction that "the discipline of a sugar plantation is exact as that of a regiment" echoed through the Louisiana sugar region where planters instituted a managerial order that resembled, Timothy Flint added, "a garrison under military discipline." Sugar planter Bennet H. Barrow recognized that and counseled his overseers, "a plantation might be considered a piece of machinery, to operate successfully, all of its parts should be uniform and exact, and the impelling force regular and steady." Barrow's advice resonated throughout the sugar country, where planters established highly disciplined and drilled work gangs that labored to supply the incessant demand of the steam-powered mill. Indeed, by the 1830s, the entire plantation compound increasingly resembled a factory, with work gangs operating in shifts, and the clock dictating, with metronomic efficiency, the time each crew was called to the mill or fields. As one New Yorker remarked, the imposing sugarhouse seemed akin to a "New England factory, with its tall, smoky chimney, and mill-roll buzzing." Reverend Robert Everest, the late chaplain to the East India Company, similarly recalled that the Louisiana sugar mills reminded him of "the tall, brick chimneys of the sugar factories" that dotted the banks of the Nile, while Charles Lanman observed that "the factory-looking sugarhouses with their towering chimneys" dominated the landscape. Within the mill, the industrialized and factorylike pace of the sugarhouse thrilled and horrified visitors during the annual grinding season when the slaves toiled to granulate and manufacture dark, rich sugar. Indeed, by pursuing prudent slave management and a sagacious division of labor, planters confidently asserted that, "free labor cannot compete, in the manufacture of sugar, with better organized slave labor."[13]

13. James Ramsay, *An Essay on the Conversion of African Slaves in the British Sugar Colonies* (London, 1784) quoted in David Barry Gaspar, "Sugar Cultivation and Slave Life in Antigua before 1800," in *Cultivation and Culture: Labor and the Shaping of Slave Life in the Americas,* ed. Ira Berlin and Philip D. Morgan (Charlottesville: University Press of Virginia, 1993), 114; Timothy Flint, *History and Geography of the*

To establish such estates, planters funneled immense resources into land, slaves, and capital improvements. Indeed, agricultural commentator Edward Forstall estimated that by 1844-1845, the total capital investment in the sugar industry stood at $60 million or approximately $78,740 per plantation. By the mid-1850s, capital expenditure had spiraled still further with medium-sized estates worth over $125,000 while the largest plantations were valued in the hundreds of thousands of dollars. As table 1 documents, even land and labor lords who wished to puncture the one hundred-hogshead ceiling had to effectively double their capital resources and to join the small coterie of elite sugar lords, estate managers needed to secure loans of between one-third and one-half of a million dollars. The local pressmen concurred, observing in the *Pioneer de l'Assomption* that "it is necessary to have great capital or immense credit" to succeed in the antebellum sugar trade. Planter Moses Liddell spoke from firsthand experience when he concluded: "if you go at sugar it will take you three years before you can procure seed or plant cane to make a full crop-you have an extensive building-you must have a steam engine and mill[,] $4500[,] expenses putting it up and keeping it in order, risk of crops, and continuous unforeseen . . . expenses that will eat up the profits . . . It is true that some very large fortunes have been realized at sugar planting but with an immense exertion and capital to commence with or a strong mind and over laborious perseverance." Those without financial reserves or ready access to credit in the New Orleans banks swiftly found themselves squeezed out as the productive thrust of the industry increasingly lay in the hands of the larger sugar masters.[14]

Drawing upon multiple sources of credit, sugar planters utilized not solely the New Orleans banks to fund capital investment, but

Mississippi Valley, 2 vols. (Cincinnati: E. H. Flint, 1833) 1:244–45; Edwin A. Davis, *Plantation Life in the Florida Parishes of Louisiana, 1836–1846: As Reflected in the Diary of Bennet H. Barrow* (New York: Columbia University Press, 1943), 409–10; A. Oatley Hall, *The Manhattner in New Orleans; or Phases of "Crescent City" Life* (New York: J. S. Redfield, 1851), 121; Rev. Robert Everest, *A Journey through the U.S. and Part of Canada* (London: J. Chapman, 1855), 107; Charles Lanman, *Adventures in the Wilds of the United States and British American Provinces* (Philadelphia: J. W. Moore, 1856), 2:209; Robert Russell, *North America, Its Agriculture and Climate: Containing Observations on the Agriculture and Climate of Canada, the United States, and the Island of Cuba* (Edinburgh: A. & C. Black, 1857), 249.

14. Forstall, *Agricultural Productions of Louisiana*, 4; Napoleonville *Le Pioneer de l'Assomption*, October 26, 1851; Moses Liddell to John R. Liddell, July 28, 1845, Liddell (Moses, St. John R., and Family) Papers, LSU.

Table 1.
Louisiana Sugar Estates in 1853, subdivided by Sugar Yield, Mean Capital Value, and Aggregate Capital Value.[1]

NUMBER OF SUGAR HOUSES	CAPACITY OF SUGAR YIELD	MEAN CAPITAL VALUE (SLAVES/ MACHINERY)	AGGREGATE CAPITAL VALUE
547	100 hhds or less	$40000	$21920000
347	100 to 200	$75000	$26025000
232	200 to 300	$90000	$20884000
132	300 to 400	$125000	$16500000
81	400 to 500	$150000	$12150000
64	500 to 600	$175000	$11200000
33	600 to 700	$200000	$6600000
14	700 to 800	$225000	$3150000
9	800 to 900	$250000	$2250000
10	900 to 1000	$275000	$2750000
6	1000 to 1100	$300000	$1800000
2	1100 to 1200	$325000	$650000
3	1200 to 2000	$350000	$1050000

1. *Hunt's Merchant's Magazine* 30 (April 1854): 500; Lewis Cecil Gray, *History of Agriculture in the Southern United States to 1860,* 2 vols. (Washington, DC: Carnegie Institute of Washington, 1933) 2:743.

their hugely valuable slave crews served as collateral to sustain debts. Although bank interest rates varied from 6 to 9 percent, specie reserves within the state grew at an advantageous 6 percent per annum from 1819 to 1861, and highly mobile and liquid capital contained within the slave population ensured financial stability and the viability of planting operations even when poor harvests or low prices afflicted the industry. To be sure, planters complained vociferously that "success in sugar as well as cotton planting is dependent on so many circumstances, that it is as much trusting to luck as betting on a throw of a dice," but the relatively mature network of property banks, such as the Consolidated Association of the Planters of Louisiana, enabled planters to mortgage their own property as collateral for specie and thus partially fulfill their growing demand for rural credit. Indeed, New Orleans's privileged position within the nation's financial structure assured that Louisiana ranked third in banking capital (after New York and Massachusetts) in 1840, and on the eve of the Civil War, state bankers still commanded fourth place nationally. Much of this capital underpinned the cotton and slavery boom of the southwestern states, but Louisiana's financial system, despite contraction in the wake of the financial panics of 1837, sufficed the capital requirements for economic expansion in the sugar parishes. As *De Bow's Review* announced toward the end of the antebellum era, the New Orleans banking system, while surely expanding both capital and credit, "has been tested and not found wanting."[15]

The soaring capital costs associated with sugar farming prompted a merger wave in the late 1840s as planters sought to enlarge and consolidate their estates by buying up the lands of smaller neighbors and purchasing both machinery and slaves for their new, large-scale sugar production. To spread the expense of sugar production, share risk, and cushion themselves from a potentially bankrupting run of bad crops, planters throughout southern Louisiana also wisely consolidated their investments by merging with other landholders to create cooperative partnerships that utilized financial advantages similar to those with greater individual assets. The multiple credit ratings that partnerships utilized when borrowing capital or purchasing supplies enabled cooperatively owned estates to wield more leverage in financial markets than could individually owned operations. Additionally, partnerships enabled smaller planters and urban investors to invest in the social prestige of planting and slaveholding without either risking their fortunes, however modest they might be, in the sugar economy. Even in the relatively wealthy Iberville and St. James Parish, as many as one third of the estates were managed as partnerships either between family members or among several commercial planters and businessmen. In practice, cooperative estates were no larger than single-unit enterprises. Partnerships in Iberville Parish produced 142 hogsheads during 1849 and 1850 in contrast to 143 hogsheads produced by single-owner estates. For those same years, cooperatively owned plantations in St. James Parish, along the Mississippi, produced 273 hogsheads to the 271 for individually

15. Frederick L. Olmsted, *A Journey in the Seaboard Slave States,* 2 vols. (1856, reprint, New York: G. P. Putnam's Sons, 1904) 2:318–19; *De Bow's Review* 25 (November 1858): 559. On banking, see George D. Green, *Finance and Economic Development in the Old South: Louisiana Banking, 1804–1861* (Stanford: Stanford University Press, 1972); Larry Schweikart, *Banking in the American South from the Age of Jackson to Reconstruction* (Baton Rouge: Louisiana State University Press, 1987), 258–59; Richard Holcombe Kilbourne Jr., *Debt, Investment, Slaves: Credit Relations in East Feliciana Parish, Louisiana, 1825–1885* (Tuscaloosa: University of Alabama Press, 1995).

owned farms. Predictably, the largest operators also featured a relatively large proportion of cooperatives as planters combined credit to purchase additional land, slaves, and the latest and most sophisticated refining equipment. Among the top 10 percent of planters in St. James Parish, for instance, four of the seven vast estates that yielded from 800 to 1,400 hogsheads in 1858 and 1859 featured multiple owners who could draw on multiple assets to sustain production at these elevated levels.[16]

Capital investment in land, labor, and agricultural technology advanced swiftly as landholders both expanded their estates and converted from animal to steam power. Indeed, sugar estates far surpassed their southern cotton and midwestern wheat neighbors in every index of plantation size and capital investment. As historian Gavin Wright indicates, farms in the rich cotton belt included an average of 130 improved acres per estate, while those in the old Northwest averaged 70 improved acres per farm. Total farm value similarly tilted in favor of the cotton South where the mean farm value of $4,370 dwarfed the estimated farm value of $2,958 for the Northwest. These figures, however, prove minuscule in comparison with those from the Louisiana sugar country, where even the smallest sugar estates outstripped their relative equals in cotton and where medium-sized operators possessed the scale and scope for economic advancement. In the relatively isolated St. Mary Parish located in the westerly Attakapas district, for instance, investment on local sugar estates advanced rapidly in the 1850s; at midcentury, 174 farmers in the parish produced sugar and owned farms with an average cash value of $23,948. These agrarians cultivated just over 40,000 acres and collectively owned $700,000 worth of farm machinery. A decade later, the profile of agriculture in St. Mary Parish proffered a transformed visage: 152 farmers now cultivated cane, though despite the reduction in total number of farmers, the number of improved acres had increased to 62,000 while the cash value had soared to $9,421,000. Machinery investment likewise had doubled from an average of $4,034 per farm to $7,612 in the space of the decade as farm managers and planters replaced the final mule- or oxen-driven mills with steam-powered technology. That said, they still continued to produce sugar in open kettles. In Ascension

16. Richard Follett and Rick Halpern, "Race and Labor in the Cane Fields: Documenting Louisiana Sugar, 1844–1917," AHRC Project Grant 16429. Exact URL available from author.

Parish, on the Mississippi River, where planters had established extensive estates in the 1830s, the sugar industry featured still greater wealth, both in land and capital. At midcentury, sugar estates in the parish featured an average of 460 acres, valued at $107,000 with almost $11,000 invested in farm machinery. By the eve of the Civil War, the average sugar estate in Ascension Parish included over 800 acres of improved farmland and well over $18,000 of machinery. The Mississippi river lords were unquestionably larger than their neighbors to the West and possessed not only steam-powered grinding facilities but also the latest vacuum technology for the production of a whiter and more valuable sugar. Bagasse burners, which allowed planters to replace wood with dried cane husks as their primary fuel, dotted the landscape; draining machines reclaimed lower portions of their land; and improved steam engines were shipped to riverside plantations from foundries in the north. Although technologically at least five to ten years behind their wealthy neighbors along the Mississippi "coast," planters in the Attakapas need only have looked to Ascension Parish to observe the future of the sugar industry unfolding before their eyes. To be sure, they may have lacked the hugely expensive vacuum machinery, but like their brethren in other parts of the sugar region, St. Mary Parish planters were not only willing to invest in machinery and alter production techniques to match the demand for quicker and more efficient grinding, but they collateralized credit and leveraged their assets to finance the agroindustrial transformation.[17]

As economic historians Ransom and Sutch have concluded, in many parts of the American South, "slaves as assets crowded physical capital out of the portfolios of southern capitalists." As such, southern farms remained technologically backward while regional economic development grew at a sluggish and disappointingly slow pace. Not only did slavery arrest economic progress on southern farms, but with their lower capital-labor and land-labor ratios, southern plantations possessed little incentive to adopt or invent

17. Gavin Wright, *The Political Economy of the Cotton South: Households, Markets, and Wealth in the Nineteenth Century* (New York: W. W. Norton, 1978), 48. U.S. Bureau of the Census, *Seventh Census of the United States,* manuscript agricultural schedules, Ascension Parish and St. Mary Parish, Louisiana, 1850; and U.S. Bureau of the Census, *Eighth Census of the United States,* manuscript agricultural schedules, Ascension and St. Mary Parish, Louisiana, 1860.

new labor-saving machinery unlike small labor-constrained north-ern farms where technical change helped achieve higher profits. Although accurate for the cotton South, climatic and crop impera-tives in the Louisiana sugar industry prompted landholders and estate managers to realize time and labor savings by using capital-intensive machinery to grind and process their crops as swiftly as possible. As such, machine investment powerfully distinguished the Louisiana sugar industry from its neighboring cotton complex in Mississippi and Alabama or the wheat culture of the Midwest. The value of implements and farm machinery to improved acreage in the free states averaged $1.60, while in the South, most cotton farmers invested $1.46 in tools for each cultivated acre. In the sugar country, by contrast, the planters spent between $17.52 and $23.27 on implements and machinery per improved acre in 1850.[18]

The sugar masters expended well over ten times the capital their brethren in cotton and wheat disbursed in implement and machine investment. Louisiana accordingly led the nation in machinery and implement investment per farm by a factor of three with almost $19 million disbursed on agricultural capital throughout Louisiana's agricultural belts. Only New York and Pennsylvania surpassed Louisiana in the total value of farm implements, though these states possessed 195,000 and 156,000 farms, respectively, while Louisiana possessed just 17,281 farms of which fewer than 10 percent pro-duced sugar. Those cane estates, however, absorbed the vast major-ity of the agricultural capital, assuring that the Louisiana sugar country was the most heavily capitalized and investment-rich agricultural region in the country and one that towered above its southern neighbors. Within the sugar country, the value of farm implements per parish frequently outstripped those of entire north-ern free-labor states. West Baton Rouge featured more mechanical capital than all of Minnesota in 1860 while Ascension Parish's 232 estates almost matched the combined total of 5,600 farms in Oregon. St. James and St. Mary dwarfed old states like Rhode Island and trumped Kansas and Nebraska combined. But despite

18. Roger Ransom and Richard Sutch, "Capitalists without Capital: The Burden of Slavery and the Impact of Emancipation," *Agricultural History* 62 (summer 1988): 133–60, esp. 138–39; Wright, *Political Economy of the Cotton South*, 52; U.S. Bureau of the Census, *Seventh Census of the United States*, manuscript agricultural schedules, Ascension Parish and St. Mary Parish, Louisiana, 1850; and U.S. Bureau of the Census, *Eighth Census of the United States*, manuscript agricultural schedules, Ascension and St. Mary Parish, Louisiana, 1860.

the states' primacy in steam engines, the lower Mississippi valley was ill prepared for sectional conflict. Beyond the riverboats that steamed their way along the rivers and bayous, the state lacked a substantial infrastructure, both for transportation and for industry. The capital-rich sugar belt drained available resources and ensured that while the regional plantation economy flourished, it did so in a relative vacuum of statewide inertia.[19]

Private investment in technology, however, increasingly left the locus of growth in the hands of the largest and wealthiest sugar masters; an elite group of approximately five hundred planters who owned over two-thirds of the slaves and available acreage in the cane world and who produced three-quarters of the regional sugar yield. These individuals possessed both the capital and credit to purchase advanced facilities for sugar producing and refining. As Frederick Law Olmsted observed, they were "among the most intelligent, enterprising, and wealthy men of business in the United States." Eager to tap the burgeoning demand for white crystalline sugar, the elite sugar masters funneled their resources into evaporation technology and clarification facilities that produced "large and brilliant crystals . . . [of] any size required by the caprice of the customer." Adapted for plantation use by Norbert Rillieux, vacuum processing transformed the manufacturing of sugar in the second half of the nineteenth century. Born the son of a white plantation owner, Rillieux was trained in the Parisian school of mechanical engineers before returning to New Orleans in 1833. He was keenly aware that the open-kettle method of sugar production was costly in terms of timber for the furnaces and hazardous for laborers on the mill floor. Rillieux's process evaporated the molten sugar in a series of sealed vacuum pans that used the heat generated by the exhaust of a steam engine rather than the direct heat of a furnace. Following the patenting of his multiple-effect vacuum pan in 1843, Rillieux's apparatus resolved the production requirements of the largest cane lords who sought to manufacture higher-grade sugars. First, the use of a steam vacuum minimized the risk of scorching or discoloring the sugar, and for planters like Maunsell White, it produced sugar of a superior quality, worth fully two cents a pound higher than the best previous sugars. Vacuum-produced sugar, White concluded, was in short, a "fancy article." Second, vacuum

19. U.S. Bureau of the Census, *Eighth Census of the United States* (New York, 1861).

pans maintained a lower average temperature than open kettles, and by evaporating the sugar in a sealed unit, producers found that the quality and quantity of the final product surpassed the caliber of sugar made by all previous methods.[20]

For the very richest planters, like Judah Benjamin, the demand for improved sugar and a whiter article necessitated the shift toward costly vacuum processing. Acquiring Rillieux's multiple-effect evaporator for use on Bellechasse Plantation, Benjamin formed a partnership with Theodore Packwood for the production of sugar with the new technology. Investing over $30,000 for this privilege, Benjamin calculated that for the largest operators the Rillieux apparatus would generate a profit of $14,531 every season over the open-kettle method of production. Not only would planters save considerably on timber and produce higher-quality sugar, Benjamin enthusiastically declared, but the new vacuum system would produce 25 percent more sugar. While some planters debated the competency of slave labor to manage the complex machinery, the Rillieux apparatus and allied vacuum evaporators proved too expensive for most planters. Despite the cost, however, over sixty-five prominent sugar cultivators pioneered these new technologies and produced an ever-increasing volume of crystalline and snowy sugar that Princeton chemist R. S. McCulloh praised as "equal to those of the best double-refined sugar of our northern refineries."[21]

Notwithstanding these episodes of expensive capital investment, almost 90 percent of planters continued to produce sugar in open kettles. Local efforts to disseminate the latest scientific findings failed because of the relative isolation of the region's cane lords and the lack of communitywide support. Moreover, the instability of Louisiana sugar and the enormous cost of taking the next

---

20. Olmsted, *Seaboard Slave States*, 671–72; James D. B. De Bow, *The Industrial Resources, Etc., of the Southern and Western States*, 3 vols. (New Orleans, 1853), 2:206. On technological improvements in sugar production, see Heitmann, *Modernization*, 8–48; Nöel Deerr, *The History of Sugar*, 2 vols. (London: Chapman & Hall, 1950), 2:561-69.

21. *De Bow's Review* 5 (February 1848): 292–93; De Bow, *Industrial Resources*, 2:206; Report of the Secretary of the Treasury, 29th Congress, 2nd Session-Senate Doc. No. 209. "Investigations in Relation to Cane Sugar: A Report of Scientific Investigations Relative to the Chemical Nature of Saccharine Substances, and the Art of Manufacturing Sugar; Made under the Direction of Professor A. D. Bache by Professor R. S. McCulloh," (Washington, DC: Ritchie & Heiss, 1847), 121; Mark Schmitz, "Economic Analysis of Antebellum Sugar Plantations in Louisiana," (Ph.D. diss., University of North Carolina at Chapel Hill: University of North Carolina Press, 1974), 39.

technological step beyond steam-powered mills and open-kettle production set the threshold too high for most individuals. Eager boosters like editor Robert Wilson of the Franklin *Planters' Banner* critiqued the apparent conservatism of the Attakapas planters toward the costly vacuum equipment, damning their aversion to "innovations and improvements." But as James De Bow readily explained, "scarcely any of the planters of Attakapas have adopted the new improvements in sugar making, which are in such progress in other Parishes." De Bow concluded that "for a great part, the estates are too small unless the machinery were less expensive."[22]

Also, planters discovered that the most expensive machinery often led to diseconomies of scale. Indeed, investment tended to favor relatively small sums of up to $3,500—significantly, that was the price of a good secondhand steam-powered sugar mill. Investment at this level effectively doubled the planters' output as these mills possessed an "excess-capacity for expansion," and once suitably equipped, planters could ultimately augment their operations, expand onto additional acreage, confident that their machinery would rapidly and efficiently mill the crop before the first killing frosts descended. It was thus the relatively simple and modestly priced steam engine, rather than the largest or latest equipment, that broke the technological bottleneck to production for most Louisiana farmers. Planters nonetheless found a strong relationship between investment in farm implements and sugar production. This remained true when investment in machinery stayed under $20,000, but those who invested in the costly vacuum facilities did not see a corresponding dollar-for-dollar increase in output. The very largest estates similarly suffered from uncompetitiveness in sugar production per hand. Indeed, for the optimal use of slave labor in 1859, preferred sugar estates contained fifty-one to one hundred slaves with each hand producing a mean of 3.4 hogsheads per annum. In contrast, the largest plantations produced only 2.1 hogsheads of sugar per hand in 1859, or almost one-third less than moderate-sized, more efficient sugar plantations. Despite the difference in scale between sugar and cotton production, Gavin Wright's essential premise for the cotton South that "[it] is not that larger slave farms were more efficient, but that there was an upper bound on the possibility of efficient expansion" appears true for sugar.

22. *Franklin Planters' Banner,* September 20, 1845; April 18, 1848.

Whereas that upper bound in the cotton fields of Mississippi and Alabama favored planters with sixteen to fifty slaves, the limit to efficient sugar farming tended to be eighty to one hundred slaves. In Terrebonne Parish, for instance, the most efficient sugar master in 1859 was Adolphe Verret, a forty-one-year-old sugar planter on Bayou Black whose seventy-nine-strong slave workforce produced 5.5 hogsheads of sugar each. A model of a highly successful though moderately large sugar planter, Adolphe Verret found that he could maximize output with moderate-sized crews and relatively inexpensive machinery. Thus, while steam-powered milling technology enabled planters to expand their acreage to match the grinding capacity of the mills, estate managers also experienced countervailing pressures toward maintaining efficient moderate-sized estates.[23]

Essentially, planters on the largest estates faced long-term managerial diseconomies of scale where the overseer and planter experienced increasing difficulty in sustaining control over their workers. In cotton farming, one overseer could maximize his managerial capacity on an estate with fifty working hands, but the highly brigaded and heavily routinized nature of sugar farming ensured that cane overseers could keep watch over slightly more workers than their compatriots in the Alabama black belt. Moreover, planters on the largest estates faced problems of control loss over their large slave crews; this was certainly not new or radically different from other nineteenth-century industries where despite the apparent advantages of growth, management difficulties tend to favor a lower optimum size. Planters faced what economists often describe as hidden-action and hidden-information cases; such scenarios address the problematic consequences of the interaction of individuals where the actions of the one (the agent) affect the interests of another (frequently his employer or principal).

23. Mark D. Schmitz, "Economies of Scale and Farm Size in the Antebellum Sugar Sector," *Journal of Economic History* 37 (December 1977): 978–79; Schmitz, "Economic Analysis of Antebellum Sugar Plantations in Louisiana," 221–26; Wright, *Political Economy of the Cotton South*, 83–85. On the upper bound to cotton farming, also see Fogel and Engerman, *Time on the Cross*, 193–96; Fogel, *Without Consent or Contract*, 73–75. Data compiled from Schedule 4, *Productions of Agriculture and Schedule 2, Slave Inhabitants, Seventh Census of the United States, Terrebonne Parish 1850* (Washington, DC, 1850); *Eighth Census of the United States, Terrebonne Parish 1860* (Washington, DC, 1860); Champomier, *Statement of Sugar Crop in 1849–1850*, 36–38; *Statement of the Sugar Crop in 1859–1860*, 28–30; Karl Joseph Menn, *The Large Slaveholders of Louisiana—1860* (New Orleans: Pelican, 1964), 413–19.

Managerial problems inevitably rise when the principal lacks key knowledge or experience that the agent possesses or when the principal simply can no longer oversee the actions of each of his agents. On antebellum sugar plantations, these two difficulties repeated themselves fairly frequently, leading to managerial difficulties and appropriate compensation systems that tempted the agents (slaves) to act in their principal's (slaveholder's) interest. It is an obvious truism that the effort and work rate of the slave (agent) had a direct impact upon the business success of the planter (principal). This, however, was complicated in hidden action scenarios by the planter's inability to perfectly monitor the slaves' work. The size of the estate, the multiplicity of tasks, and the overseer's marginal (at best) advantage in planting expertise all complicated the task of monitoring the slaves' work in the field and mill. Regimented gang labor partially resolved the planters' dilemma, but the planters could not easily resolve the central problem of control. While overseers faced that dilemma, they also faced hidden-information problems in which the principal must motivate the agent to put his/her technical expertise or information into action, even if both participants are aware that the principal lacks the detailed knowledge or managerial capacity to judge whether the agent is working in his/her interest. In the sugar country, most masters and overseers knew when slaves were openly acting against the planters' affairs, yet they nonetheless confronted the typical principal's predicament of motivating those individuals with expertise to work for them. Planters certainly introduced incentives to encourage their often skilled workers to labor long hours and regimented their crews into disciplined gangs, but the combination of control loss, the overseers' incapacity to monitor everyone's work, and the limits on cash payment in a slave system ensured that managers on the largest estates faced substantially more problems than those on smaller estates. Slaveholders could appoint assistant overseers, but the managerial dynamics of the sugar world led toward sustained efficiency among midrange plantations.[24]

24. Jacob Metzer, "Rational Management, Modern Business Practices, and Economies of Scale in the Ante-Bellum Southern Plantations," *Explorations in Economic History* 12 (April 1975): 144; Sidney Pollard, *The Genesis of Modern Management: A Study of the Industrial Revolution in Great Britain* (Cambridge: Harvard University Press, 1965), 10; Oliver E. Williamson, "Hierarchical Control and Optimum Firm Size," *Journal of Political Economy* 75 (April 1967): 124; Peter A.

Despite boastful rhetoric from local pressmen, the Louisiana sugar industry never evolved during the antebellum years into a technically advanced refinery-based trade producing high-grade sugars. Instead, the region's planters produced brown plantation-grade sugar that found a ready market in the West. To be sure, they exploited the growing demand for sucrose, profited from technical gains in productivity, utilized sophisticated financial instruments, and thrived behind the lofty walls of federal tariff protection. Yet for all the industry's relative modernity, planters shied away from the most costly technical innovations. The volatility of the sugar economy and the Gulf climate unquestionably bred conservatism among the planter class, and their managerial difficulties with the largest estates led many to favor steam-powered milling with open kettles over the expensive new vacuum pans. The insecurity of tariff protection and growing Cuban competition worried planters still further and led to reservations over high-level investment. As Moses Liddell well understood, the never-ending costs of sugar production drained the planters' coffers, and plantation debts, or the impending threat of them, constrained the economic vision of all save the largest sugar masters. Added to this, antebellum planters remained too self-interested to invest in shared refining equipment for their estates or move toward centralized milling. As two decades of scholarship has indicated, planters could modernize and embrace the market economy, but their identity as slaveholders ultimately defined progress in the slave states. Above all, dogged independence and steadfast individualism derived from the specific cultural values of the slave-plantation complex. The planters' social ethic and economic ideology exalted the slaveholders' authority, celebrated their power, and emphasized personal liberty. Such values remained anchored to the plantation world and to the planters' role as slaveholders and labor lords. Most of all, it encouraged

---

Coclanis, "How the Low Country Was Taken to Task: Slave-Labor Organization in Coastal South Carolina and Georgia," in *Slavery, Secession, and Southern History,* ed. Robert Louis Paquette and Louis A. Ferleger (Charlottesville: University Press of Virginia, 2000), 66-67. On the hidden action scenarios, see David E. M. Sappington, "Incentives in Principal-Agent Relationships," *Journal of Economic Perspectives* 5 (spring 1991): 45-66; John W. Pratt and Richard J. Zeckhauser, "Principals and Agents: An Overview," in *Principals and Agents: The Structure of Business,* ed. John W. Pratt and Richard J. Zeckhauser (Boston: Harvard Business School Press, 1985), 1–35; Robert Gibbons, "Incentives in Organizations," *Journal of Economic Perspectives* 12 (fall 1998): 117, 121–23.

individualistic notions of progress and a regional myopia to collective investment or collaboration for the communal good. The nascent Agriculturists' and Mechanics Association and the University of Louisiana failed to secure adequate support, and institutions necessary for technical progress expired due to planter intransigence. Associations committed to statewide commercial progress similarly failed for want of support and public interest. To be sure, the vast cost of taking the next technical step toward vacuum processing checked the widespread dissemination of the latest technology, but the institutions and commercial interest of the sugar planters toward innovation and association proved wanting. Self-absorbed commercialism accordingly reigned triumphant, but it was ultimately contained within the privacy of the planters' estates and sugarhouses.[25]

Ever astute, New Yorker Frederick Law Olmsted accurately pointed to the overwhelming focus on private economic progress when he observed in his travels through central Louisiana that "there was certainly progress and improvement at the South . . . but it was much more limited, and less calculated upon than at the North." Olmsted's observations proved harsh, though his central argument that there was "no *atmosphere* of progress and improvement" proved tellingly accurate. As he saw it, "there was a constant electric current of progress" in the North, but in the South, "every second man was a non-conductor and broke the chain." Characteristically, the perceptive visitor put his finger on the central problem. "Individuals at the South," he added, "were enterprising, but they could only move themselves." Herein lay Louisiana's greatest flaw—sugar planters invested thousands of dollars in the New Orleans slave market and in a new steam-powered mills, yet their vision of economic growth remained bound to their estates, their own profit margins, and private, independent enterprise. It was a philosophy perfectly suited to plantation agriculture but one ill-suited for the capital-rich technical age upon which Louisiana sugar was about to embark.[26]

---

25. On the state university and planter diffidence to collective policy, see Follett, "'Give to the Labor of America, the Market of America,'" 117–47; Heitmann, *Modernization,* 40–48; Walter L. Fleming, *Louisiana State University, 1860–1986* (Baton Rouge: Louisiana State University Press, 1936), 3–20.

26. Olmsted, *Seaboard Slave States,* 274–75.

Building Networks of Knowledge

Henry Merrell and Textile Manufacturing in the Antebellum South

MICHELE GILLESPIE

HISTORIANS TEND TO RELEGATE THE DEVELOPMENT OF ANTEBELLUM southern manufacturing to second-class status, insisting on the primacy of the cotton economy and plantation agriculture in their continual quest to explain the region's uniqueness. As a result, they have paid scant attention to the impact of technological transformation on manufacturing, including the antebellum southern textile industry. Fred Bateman and Thomas Weiss determined that the paucity of sustained manufacturing in the South compared with the North and West reflected planter preference for acting as individualistic slaveholding agriculturalists intent on operating beyond the reach of legal constraints. But that position does not adequately account for change over time and the push and pull on planter capitalists to move into the manufacturing realm by the late antebellum era.[1] Southern planters, entrepreneurs, and statesmen paid close attention to the spread of cotton manufacturing from England to the northeastern United States. They gauged the political and economic possibilities of introducing mills to their own region as closely as they gauged the tariffs. These planters subsequently invested in cotton manufactories, especially during the weakened economy of the 1840s, in what arguably marked the real beginnings of southern manufacturing. Meanwhile, northern mechanics and entrepreneurs eager to capitalize on cotton manufacturing opportunities in the

1. Fred Bateman and Thomas Weiss, *A Deplorable Scarcity: The Failure of Industrialization in the Slave Economy* (Chapel Hill: University of North Carolina Press, 1980), 26-37, 41–46, 58, 157–62.

South became critical bridges for the introduction of cotton manu-facturing's attendant new technologies to the region.

In Georgia, as in most other southern states, a few capitalists had begun to invest seriously in internal improvements by the late 1820s in response to strong internal migration by farmers and planters intent on getting their new upland cotton to market. Private banks supported ventures for steamboat and later railroad transport, received state sanction through regulation, and ultimately benefited from the booming new economy.[2] By the late 1830s, homegrown capitalists were building textile mills in the belief that Georgia's strong economy coupled with political divisiveness between the North and the South augured well for domestic manufactures. The earliest factories were simple, small efforts, committed to the pro-duction of rough cloth. But by 1840 the Empire State sported more than a dozen cotton mills, employing nearly one thousand hands, with capital investments totaling more than half a million dollars. Those numbers had more than doubled by 1850. Thus the South was eagerly duplicating the New England model.[3]

Knowledge of cotton technology and its rapid transformation was at a premium in the region as a result of these developments, and new technology played a critical role in advancing southern indus-trialization. Gavin Wright has argued that the antebellum South experienced no "spectacular technological breakthroughs" during this era because it suffered from labor market isolation and lacked a "southern technical community." Hence southern capitalists who invested in cotton manufacturing had to rely mostly on secondhand knowledge from the larger Atlantic community to build and operate their mills.[4] They left the work of improving upon technologies to their more experienced counterparts in the northeast and Britain.[5]

Scotsman James Montgomery, in his well-known assessment of the state of cotton manufacturing in America, observed that

2. Kenneth Coleman, *A History of Georgia* (Athens: University of Georgia Press, 1977), 153–54.

3. Reported in "Industry and Commerce" manuscript, Georgia Writers Project, WPA, Box 57, File 7, Hargrett Library, University of Georgia, Athens, Georgia.

4. Gavin Wright, *Old South, New South: Revolutions in the Southern Economy Since the Civil War* (New York: Basic Books, 1986), 62, 79, 125, 157.

5. See especially David Jeremy, "British Textile Transmission to the U.S.: The Philadelphia Region Experience, 1770–1820," *Business History Review* 47 (spring 1973): 24–52; David Jeremy, *Transatlantic Industrial Revolution: The Diffusion of Textile Technologies between Britain and America, 1790–1830* (Cambridge: MIT Press, 1981).

Americans were proving impressive rivals with Great Britain for the top spot in textile production. He also noted the potential for serious competition between the northern and southern regions of the United States. His trip through the country convinced him that America's great water sources and ample cotton cultivation gave the United States "advantages which no other nation enjoys." Great Britain's continued dominance in the industry, he thought, depended almost solely on its improvements in machinery to reduce production costs. Montgomery's reasoning was understood by virtually all the key players in the international textile game, including those in the South. Meanwhile, the Scotsman, after touring the region, predicted particularly successful ventures for southern manufacturers, who had special advantages because they cultivated their own cotton, saving them the expense of shipping and transport, and perhaps the advantage of cheaper labor, too, if slaves were employed at the looms. Still, Montgomery concluded, the North's abundance of skilled mechanics, and their relative absence in the South, would make the two regions competitive with each other and give the nod to the North for many years to come.[6]

Here Montgomery understood a key point about relative advantages between northern and southern industrialization. Spreading technological knowledge about innovations in textile manufacturing was critical to the development of the antebellum textile industry in the South. This essay argues that personal networks were instrumental in that dissemination and could, in some cases, obviate the multiple social, economic, and cultural barriers that discouraged, and sometimes prevented, industrialization. This essay highlights the experience and perspective of Henry Merrell (1816-1883), a talented mechanic and mill manager who became a Georgia and Arkansas cotton manufacturer. His life and work marked his commitment to building networks of knowledge within and across regional lines to encourage technological innovation in cotton manufacturing. His networks evolved out of multiple kinds of relationships, many of which economic historians have tended to overlook. Merrell's communities of kinship, friendship, faith, politics, ideology, professional culture, and consumption played critical and

6. James Montgomery, *The Cotton Manufacture of the United States of America, Contrasted and Compared with That of Great Britain*, (New York: D. Appleton & Co., 1841), 188-94.

often overlapping roles in advancing knowledge about new technology in the southern cotton industry.[7]

In 1848, *De Bow's Review* touted Henry Merrell as "one of the shrewdest Yankee manufacturers" in the South. At midcareer, Merrell had used his profits from the sale of the Mars Hill factory in Athens, Georgia, to build a new cotton factory on the Oconee River in nearby Greene County. In fact, he believed, as did *De Bow's Review,* that he would soon "retire with a fortune."[8] Unfortunately, his decision not to sink significant capital into the best new machinery available, coupled with a rise in competition for cotton goods markets sent him tumbling into bankruptcy. These hard times postponed his retirement indefinitely. Merrell spent many years extricating himself from his debt before traveling to Arkansas to build a mill village from scratch in 1856.

Merrell may have been shrewd, but he was not shrewd enough to avoid financial disaster in the late 1840s. His back-to-back failures at two enterprises, however, cemented his long-held belief in the importance of four key ingredients for successful cotton manufacturing in the South: First, he believed that capitalists, managers, mechanics, and operatives alike must commit themselves to a virtuous work life fueled by their Protestant faith. Second, he held that investors and managers must purchase only top-notch, up-to-date machinery, to be secured from the best machine shops in the Northeast regardless of cost. Third, he was convinced that manufacturers and mechanics must build communities in which they exchange technical information to stay abreast of the latest developments to ensure their success. Finally, he contended that manufacturers in southern states must support one another by sharing knowledge and resources in an ever-widening world market rather than compete against one another locally.

Merrell had determined the basics of his four-part "Southern Manufacturing System," as he labeled it, quite early in his career. Upon his father's premature death, Merrell had entered an apprenticeship,

7. James L. Skinner III discovered Henry Merrell's unpublished autobiographical writings in his parents' home in Roswell, Georgia, in 1981. He skillfully and substantively edited Merrell's two journals over the next ten years, publishing them in a single volume, *The Autobiography of Henry Merrell: Industrial Missionary to the South* (Athens: University of Georgia Press, 1991).

8. "Industry of the Southern and Western States," *De Bow's Review, Agricultural, Commercial, Industrial Progress and Resources* 6 (October–November 1848): 203.

training to be a machinist and textile manufacturer at the Oneida Factory in Whitesboro, New York. In 1839, at the age of twenty-two, he headed south in search of opportunity. What he lacked in capital he believed he made up for in mechanical know-how and hands-on experience, which he expected to serve him well in a region he considered essentially devoid of manufacturing knowledge. Upon reaching Lexington, Kentucky, he met an old family friend, a devout Presbyterian like himself, who told Merrell about a group of Presbyterian planter families relocating from the Georgia coast to the healthy mountains to build a mill village based on the Protestant work ethic.

Merrell, who had come of age in the burnt-over district of upstate New York, a region hit hard by Protestant revivalism during the Second Great Awakening, liked what he heard about these investors. He sincerely believed that good habits of industry ensured deep Protestant faith and vice versa. He also felt such practices of work and faith helped equalize differences between rich and poor, although he would abandon this particular aspect of his "southern system" over the course of his adult life, especially several decades later, when he moved to Arkansas and lived among "the wild Arabs," as he called the local frontiers people. But in 1839, the Roswell experiment very much appealed to him. He promptly wrote Barrington King, one of the community's founders and head of the mills, requesting a post. King welcomed Merrell as his assistant superintendent. Hence Merrell's tightly bound Presbyterian community of kith and kin shaped the kinds of choices this young mechanic was willing to entertain at the very outset of his career.[9]

Although King had titled himself manager of the Roswell Manufacturing Company, Merrell acted as de facto operating manager from the moment of his arrival and over the next six years, he handled the engineering and mechanical departments, as well as the workers, the purchase of cotton, and the marketing of the yarn and cloth. His title remained the same throughout this time, and eventually he grew disappointed with his annual salary of $1,000, especially given the success of this enterprise under his watchful eye as measured by the escalating profits of the stockholders. Indeed, he was so committed to hands-on management that he claimed he wore out a pair of shoes every two weeks during this period. While

9. Skinner, ed., *Autobiography*, 141, 154–55, 168.

at Roswell, he married a middle-class farmer's daughter from this Presbyterian community and the couple started a family of their own. This set of circumstances compelled Merrell to leave King's employ in 1845 in search of a more profitable career.[10]

Merrell immediately purchased the aged Mars Hill manufactory near Athens, Georgia, which had not been profitable for a number of years. Yet six short months later, he managed to quadruple his investment by selling the newly renovated operation to the Curtwright Manufacturing Company. This same company, impressed with his business and mechanical acumen, hired him to run another of their factories on the Oconee River in nearby Greene County. This latter mill's directors also persuaded him to establish a steam-powered textile mill in the failing Greene County seat of Greensboro.

At this stage of his career, Merrell's talents as a cotton manufacturer were being nationally recognized. Not only was he hallowed as "a shrewd Yankee" in *De Bow's Review*, but his cotton products were earning him numerous accolades. His Curtwright cotton twist won "best and finest" at a national fair in Washington, D. C., while his cotton yarn was judged "best specimen" at the American Institute Fair in New York City, significant feats in a section of the country known for the coarseness of its threads and yarns. He also penned a series of articles on the state of southern manufacturing for journals and newspapers, including the *Southern Recorder* in 1847 and the *New York Journal of Commerce* in 1858, testament to his belief in the importance of establishing networks of knowledge.[11]

Despite Merrell's sterling national reputation, his Greene County manufactory and steam mill proved utter failures, which he later admitted were at least in part of his own making. After taking many years to extricate himself from substantial debt, he headed west in 1856, buying suitable land on a powerful river in a remote section of southwest Arkansas. Here he built not only a textile manufactory but a complete mill village that quickly secured him considerable profit and prestige in the last years before the Civil War and continued to do so until his death in 1883.[12]

Merrell wrote one account of his life shortly before the war and another late in his life, but his ideas about technology and cotton manufacturing did not change. In the manuscripts, written for his

10. Ibid., 133, 141–42.
11. Ibid., 131, xxx.
12. Ibid., 133, xxiv–xxv.

descendants with no expectation that they would ever be published, he contended that he had formulated his "Southern Manufacturing System" at the completion of his apprenticeship and even before he had begun traveling South in search of work. It was not his sole intention to create networks of knowledge to secure more profits for himself and for the region. He believed that combining a Protestant work ethic with a well-run, efficient manufacturing system would make two important contributions to southern society. First, successful southern mills would help reconcile northern and southern interests through their shared embrace of industrialization. Second, mills run by Protestants would "bring forward the low class of white people [in the South] and make men [*sic*] of them."[13] Merrell considered his own work ethic and his work skills outward manifestations not only of his godliness, but of his manhood, too. As a northerner making his livelihood in the South, and as a self-styled "industrial missionary" in an agricultural slaveholding society, Merrell would find his dearly held beliefs challenged multiple times in the years ahead.

As a youth, Merrell had internalized the mission-oriented Protestantism of middle-class manufacturers and shopkeepers in upstate New York. Their faith was a response to the growth of a new market economy as well as their engagement in the revival movement begun by Charles G. Finney that launched the Second Great Awakening.[14] Merrell believed he had a moral obligation to foster both the Protestant faith and a work ethic among the working class, and to his mind nowhere was this more necessary than in the South. Therefore, Merrell was drawn to his first post precisely because of its mission. He believed the founding families of Roswell were committed to running the mill and their new community based on the same Christian values he espoused. He presumed their work would offer a shining example of a successful business enterprise run with moral integrity, thereby "form[ing] an era in which others [could make] a fresh start" too.[15] Merrell remained proud of the Roswell

---

13. Ibid., xxii–xxiii.
14. The literature on this development is voluminous. For early formative works see Paul Johnson, *The Shopkeeper's Millennium: Society and Revivals in Rochester, New York, 1815–1837* (New York: Hill & Wang, 1978); and Anthony F. C. Wallace, *Rockdale: The Growth of an American Village in the Early Industrial Revolution* (New York: W. W. Norton, 1972).
15. Skinner, ed., *Autobiography*, 172.

enterprise and his role in it throughout his life. Not only did the company prove successful, indeed it "grew fat upon that which impoverished the planter: to wit the low price of cotton," but he believed the owners "did not lose sight of the moral, educational and religious designs of the original settlement."[16]

Merrell was not alone in his insistence on moral uplift among his southern hands, of course. Other northerners, including William Gregg, founder of the textile mill town of Graniteville, South Carolina, and Daniel Pratt, founder of the manufacturing village of Prattville, Alabama, explicitly modeled their enterprises after the "Rhode Island" system crafted by paternalistic owners and board members eager for social reform and a reliable workforce.[17]

Merrell also thought that his introduction of modern technology to the infant manufacturing business in Georgia was an important mission in and of itself. He felt his understanding of cutting-edge machinery would have a "civilizing" influence that reached beyond the Roswell community. Looking back later, he wrote: "Behold me then confirmed in my office, easy about my future in society, and rapidly bringing my machinery and my wild Arab hands into a state of organization."[18]

Unfortunately Merrell proved "only tolerably successful" at persuading local white men to learn mechanical skills and convert to his idealized work ethic. While he acknowledged that poor white men had enough "natural powers of mind" to be good mechanics, or at least operatives, their "lack of early education and religious training diluted and vitiated the whole," and he did not find them amenable to embracing that education as adults. Locals, Merrell concluded after several years of experience, were almost always useless and unredeemable because they had no scruples, were undisciplined and unfaithful, and could not be trusted.[19] To his chagrin, virtually all his efforts at training hands for all four of his southern manufacturing efforts proved equally frustrating.

At the Curtwright textile mill on the Oconee River in Greene County some years later, Merrell could not find a sufficient number of locals able to keep the machines running, which he blamed in

16. Ibid., 141.
17. Laurence Shore, *Southern Capitalists: The Ideological Leadership of an Elite, 1832–1885* (Chapel Hill: University of North Carolina Press, 1986), 30-33.
18. Skinner, ed., *Autobiography*, 154.
19. Ibid., 169–70.

part on the unhealthy location, where sickness abounded. The best went elsewhere, "leaving me only indifferent families and some of bad character." Nor could he hire slaves. "[N]o owner of Negroes would be likely to have hired us Negro hands to work at a place with such a reputation for bad morals & bad health." As a last resort, he went to New York City, where he secured "a drove of Irish immigrants" who assured him they were Protestants. He later discovered to his great dismay that many of "the girls" were not only Catholic but immoral.[20] It is interesting to note that Merrell never analyzed the efficacy of gender in his hiring practices or in his analysis of his hands and seemed committed only to hiring the cheapest and most available labor force he could locate.[21]

If the variety of hands he managed to employ—men, women, and children, both native and northern-born—were not the best group from which to build a network of Protestant mechanics, he received equally unenthusiastic responses from the other end of the social spectrum. Southern white men of the middling and planter classes, he discovered, refused to set a good example, because "for them to labor was disreputable." He was amazed to learn that many well-to-do people, especially the Roswell families who had migrated from the low country, had never seen a white man actually do physical work, at least until they met Merrell. "Even the overseers and Negro drivers rode horses and were exempt from labor," they told him. At first, Merrell responded to this culture with contempt, playing "the blunt mechanic" by purposefully rolling up his sleeves and toiling in the open whenever elite young men and women rode past. He was persuaded to stop playing the virtuous workman only by the insistence of his southern middle-class wife that he pursue the habits of his adopted region. The larger agricultural slaveholding world in which Merrell found himself, while not averse to manufacturing, refused to give much cultural ground, while workers objected to his brand of Protestant paternalism. The southern patriarchy insisted on respect for its hierarchical social order. If Merrell was to construct new networks of

20. Ibid., 197.
21. For a discussion on gender and mill hands see Michele Gillespie, "'To Harden a Lady's Hand': Gender Politics, Racial Realities, and Women Millworkers in Antebellum Georgia," in *Neither Lady nor Slave: Working Women of the Old South,* ed. Susanna Delfino and Michele Gillespie (Chapel Hill: University of North Carolina Press, 2002).

knowledge, he was going to have to work within this system. That would prove to be a challenge.[22]

Merrell believed good training would change mind-sets. He set his sights on molding several of the King sons, the children of the second-generation planter behind the Roswell venture, into manager-manufacturers. He had special hopes that Barrington King's oldest son, Charles, who had a special aptitude for mechanics, would embrace this career path. Much to Merrell's disappointment, the young man elected to study theology and became a minister instead. His career, though not lucrative, was far more suitable given the low-country planter origins of this particular social group.[23]

Merrell believed ethics and technology went hand in hand in creating successful southern manufactories. Roswell's greatest achievements, he thought, were "its management on humane and Christian principles" and "its machinery of the latest improvements." While at the Oneida factory as "a cadet in a manufacturing enterprise," Merrell had observed closely the sons of agent William Walcott. The two were a study in opposites: William Jr. proved a genius experimenter and inventor while Benjamin was "a man of talent rather than genius, of dignity and smartness, of strict discipline and sound maxims." Merrell decided that while William would most likely please the stockholders in the short term, it was Benjamin whom he should emulate, because he saw the big picture. He was "a practical man, taking care always to avail himself of improvements as fast as their economy was demonstrated." Benjamin's example had taught Merrell to seek out the best machinery throughout his career, with the exception of his Greene County enterprises, where, in the absence of adequate capital, he ignored his own advice and cut corners at his own peril.[24]

From the very start of his Georgia career Merrell's Roswell employers entrusted him with the purchase of necessary machinery and improvements. He recommended that the northern-based company of Rogers, Ketchum and Grosvenor construct the Roswell works. (In fact the first textile mill built by the Roswell Manufacturing Company was based entirely on Thomas Rogers's drawings).[25]

22. Ibid., 166, 167.
23. Ibid., 168.
24. Ibid., 172, 85–87.
25. Ibid., 135; Dumas Malone, ed., *Dictionary of American Biography* (New York: Charles Scribner's Sons, 1936), 16:112–13.

Merrell traveled to the machine works in Paterson, New Jersey, to observe and learn the mechanics of the new equipment and then supervised its loading and stowing aboard the *Milledgeville* in New York. He then journeyed with his valuable cargo, the only passenger aboard the ship. Upon arriving in Savannah, he had every intention of carefully supervising the unloading of the equipment, but the heat was so intense that he disappointed himself by "fall[ing] into the Southern way about the thing at last, and [left] the hauling of the machinery, for better or worse, to the stevedore and his Negro men, while I consulted my own comfort over . . . iced refreshments" supplied by an elderly black woman in an old man's hat who told him to call her "Momma." Although Merrell supplied many observations about social and cultural differences between the North and South throughout his accounts, he conveyed himself slipping into southern ways that belied his good habits of industry with great reluctance, and usually as a manifestation of a moment of weak character, as in the above case. Mostly he portrayed himself as what could best be described as "an industrial missionary." "Behold," he intoned when describing how his introduction of the latest machinery along with his ability to turn hired hands into principled workers had brought tremendous profits to the Roswell investors.[26]

In Merrell's mind, partnering technology and Christianity created a powerful force for an industrial transformation that brought great good to all parties. As an industrial missionary, Merrell believed he could help southerners learn about and use the latest technology available in England and the Northeast. "[I]n this race for improvement," northern men, and especially New England men, had the decided advantage in the late 1830s. Improvements traveled slowly from the North, and few improvements emerged from within the South. In his estimation, southerners were adept managers of manual labor (meaning slave labor to Merrell), but northerners were best at creating useful contrivances and machinery. Investing capital in southern manufacturing given these regional predilections necessitated the advice of a talented mechanic with extensive northern networks, someone like himself, to ensure profits.[27] While Merrell was expressing a self-serving logic here, there is truth in his underlying point.

26. Skinner, ed., *Autobiography*, 135, 137, 154.
27. Ibid., 171–72.

Merrell also recognized that textile manufacturing in the United States was changing especially rapidly in the 1840s and 1850s. "Mechanical improvements came in like a yearly revolution," he noted. In an increasingly competitive market for goods, everyone understood that the use of the fastest, most efficient machinery offered the best chance of success. Although his acquisition of the top-flight Rogers machinery had assured early profits at Roswell, within five years he was convinced these works had grown out-dated. They were old and plodding and were lowering yearly prof-its despite increased labor and economy. Meanwhile, the spread of the railroads into the piedmont was bringing even cheaper goods from more distant markets to the upcountry, eliminating the local advantage. Roswell was not alone in facing these difficulties. Virtually all piedmont Georgia factories established around 1840 suffered these same problems. Thus, Merrell observed, it was actu-ally a good thing when factories accidentally burned to the ground—because this necessitated the purchase and installation of new machinery.[28]

About this time, Merrell's cotton yarn received the silver medal as best specimen at the American Institute at New York. In an espe-cially self-critical reflection, he observed:

> Right here we ought to have stopped and re-juvenated the old Roswell factory; or, what would have been better, we ought to have adopted genuine improvements as they appeared, so keeping our Works up with the times. We did from time to time increase our works, which is one way, but not always the best way, to diminish the cost of production; but all we did was over & over again in the old way. Herein I erred. My practical objection to ingenious men had gradually and imperceptibly taken in my head against ingenious things, finally amounting to a chronic aversion against things new and improved, & a love for what was older. In a word, my usefulness as a manufacturer was drawing to an end, until such time as I should be brought by reverses and losses to set myself right by a review of the causes leading thereto.[29]

Merrell had turned his back on his own commitment to pursuing improvements upon leaving Roswell and purchasing his first factory,

---

28. Ibid., 173, 172.
29. Ibid.

Mars Hill, in Clarke County, "made up of scraps of machinery picked up here and there, & some of it so very old as to have belonged to the first factory ever started in the state, so far back as 1812. Here was a perfect museum of old things for me to flourish among, at my own private cost!" The gifted mechanic made minimal improvements and quickly trained his new hands. Six months later, he managed to sell the factory for four times its purchase price. The lightning speed of his success surely must have reinforced his decision to make do with second-class equipment when capital was limited. Later in life he concluded that his reliance on bad machinery at this stage of his career should have been his undoing, and he wished it had been. Instead, it would take him several more years and the loss of his personal fortune to learn his lesson.[30]

Merrell took his easily secured profits from Mars Hill and accepted a position as agent of a new company, in which he himself was one of the investors. He had no choice about the location of this new factory, sixteen miles south of Greensboro, on the Long Shoals of the Oconee River. Upon completing construction for this doomed enterprise, he lacked enough capital to buy the best machinery for spinning, "so I was fain to content myself and risk my reputation on low-priced machinery of New England manufacture." The machinery proved virtually worthless, and he eventually considered himself taken in by the manufacturer, a New Englander, whom Merrell assumed was "an abolitionist deliberately selling a southerner" bad equipment. "I ought to have known better than to trust them [New England manufacturers] at all."[31] This episode begs the question of how often northern manufacturers sold southerners substandard machinery, either because of ideological beliefs, especially abolitionist ones, or simply because they had yet to conceptualize the long-term value of a national reputation for fine machinery.

But even the latest technology could not ensure success when government policy undercut commercial endeavors. Not surprisingly the combination of outdated, substandard machinery coupled with the impact of the tariff of 1846 seriously damaged the businesses of many local manufacturers, not just Merrell's, in the late 1840s. "And it finally became manifest," explained Merrell, "that we had too many factories in the State of Georgia for the trade of our

30. Ibid., 173.
31. Ibid., 195, 197.

own Country, & too many for the floating population of factory hands, & that only those Factories which had the most improved machinery in the place of manual labor could hope to succeed—the rest must go under." Cotton manufacturing had changed profoundly in only five years' time. Labor had been cheap and relatively abundant and therefore had compensated for aging machinery. Cotton was cheap, so much so that perfect machinery that prevented wasting raw material seemed unnecessary. But labor, cotton, and oil prices had risen several hundred percent in the intervening years, and Merrell "began to regret bitterly that I had set my face against the improvements so far as to go on the old way instead of keeping up with the times. In the Long Shoals Factory I had actually gone backwards, and adopted new machinery of the style ten years gone by. And I had been fool enough to make a virtue of my conservatism in so doing!"[32] Merrell believed that the best improvements should override all contrary government policies.

Meanwhile, one of his protégés, James King, son of Barrington King, had spent some time in the North learning the textile mill business. He had witnessed the rapid improvements that came with the tariff of 1846 and listened to mechanics discussing top machinery. King returned to Roswell intent upon incorporating this new knowledge into the newest factory, which ultimately assured the Roswell Manufacturing Company substantial profits throughout these tough times and well into the 1850s. Merrell, who had encouraged James King to always secure the best machinery and to study the latest innovations in the Northeast, must have found his failure to follow his own advice a particularly bitter pill. Indeed he recognized in hindsight how his unwillingness to reach beyond his local community at this time had hurt him badly. "A residence of six months at the North would [have] set me right, but the plodding round of my old business and the growing necessity for hard work and close management blinded me quite."[33] It stands to reason that a more fully developed regional technical community might have rescued Merrell and many other manufacturers from this tight spot.

Meanwhile, Merrell had been invited to establish a cotton-spinning factory on the Oconee River near Greensboro. It would

32. Ibid., 198–99.
33. Ibid.

also prove unsuccessful. A stockholder provided inferior bricks, and Merrell lacked adequate funds for machinery, making do with "that which was low-priced, although at the same time I was able to adopt some but not all of the current improvements." This time he bought his machinery from a Mr. Leonard of the Mattewan Machine Company of New York City. Although it was considerably cheaper than most other machinery on the market, Leonard had long been known to be "a sharper," and the machinery Merrell purchased proved a disaster. "I was ruined [by it]," he concluded. Coming so quickly, one after the other, these experiences persuaded him once and for all that investing in the best machinery was always the shrewdest strategy. When he began construction on his Arkansas mills in 1856, he traveled north to Cincinnati to buy heavy equipment for milling that could be shipped relatively easily and cheaply, but he insisted on purchasing his fine machinery for carding and spinning from Charles Danforth in Paterson, New Jersey, a far more costly decision given not only the higher expense of the machinery itself but the additional cost of shipping. He had no regrets at the time or thereafter. "I scarcely would have had the machinery of any other builder as a gift."[34]

Danforth himself had been a factory lad once and had spent much of his career improving on carding and spinning machines. Merrell was surprised that his early "great genius" had gone unrecognized until Danforth switched to railroads, making a great fortune, at which time people came to value his textile machinery, too. Danforth had actually advised Merrell to use his machinery at the Greensboro factory. "I thought he was talking in his own interest. Had I been governed by his advice, I would have been a wealthy man still living in Georgia," admitted an older and wiser Merrell.[35]

Although Merrell had purchased the best machinery available for his Pike County, Arkansas, enterprise, and while it performed exceptionally well for him for many years, his remote location and its uniqueness challenged him as a manufacturer and an innovator. It was a big jump from the sluggish rivers of Georgia to the powerful waters of Muddy Creek, and he had to apply sophisticated engineering knowledge to construct dam girders strong enough to withstand the considerable water pressure. This was no small feat,

34. Ibid., 199, 242, 241.
35. Ibid.

and the fact that he accomplished it led people to proclaim him "the man who dammed up the 'little' Missouri River."[36]

Nor could Merrell rely on the traditional breast wheel or over-shot wheel to power his mill given the extreme changes in river conditions and the accumulation of gravel.[37] This forced him to design his own, a version of the re-action iron wheel, because he was hundreds of miles from any machine shops or foundries. Likewise Merrell had to handle all his own repairs. In Georgia, by contrast, Merrell had had access to a number of small foundries as well as growing heavy industry available in the Macon area by the 1850s, including Robert Findlay's impressive ironworks.[38] Merrell was proud of his ability to adapt and make do in Arkansas:

> Remote from foundries and machine shops, with only bungling self-taught mechanics about me, I was all the time at my wit's-end to get along, but at the same time always confident of being able to meet any emergency likely to arise, & generally by fore-casting already prepared for accidents before they took place. The story of all my contrivances in order to get on under these circumstances would make an engineering book by itself, much after the manner of Robinson Crusoe.[39]

Merrell was clearly impressed with his own accomplishments in Arkansas, painting himself as the single competent among a country of ignoramuses. In this sense, however, Merrell does deserve credit in the history of southern textile manufacturing for being an ambitious innovator. A talented mechanic and engineer, he had had the wherewithal to adapt the technical knowledge he had acquired in upstate New York to his Georgia mills, and with even greater success, to his Arkansas effort. Technological transformation and innovation in the nineteenth-century Atlantic world were marked by the ability to "borrow" technologies from other places and

36. Ibid., 255, 257, "I have recently read in some engineering book directions for obtaining durable oak timber. The writer says to go to low grounds and rich lands to select your timber [for the water wheel shaft]. His advice does not accord with my experience in the Southern climate. The most durable white oak timbers will be found growing on an elevation in an old field of poor worn out soil."

37. Ibid.

38. Robert S. Davis, *Cotton, Fire, and Dreams: The Robert Findlay Iron Works and Heavy Industry in Macon, Georgia, 1839–1912* (Macon, GA: Mercer University Press, 1998).

39. Skinner, ed., *Autobiography*, 259.

"innovate" enough with these ideas to make them work in new locales and under different circumstances, as on the Muddy River in Arkansas. The relative absence of such exemplars of innovation in the history of the antebellum South does not in and of itself mean they did not exist. Instead, as the example of Henry Merrell suggests, such innovators were indeed present, though in most cases not in such great numbers and in close enough proximity to each other to enable historians to document their innovations and transmissions.

While Merrell had some talents as an innovator, his long-term success can best be attributed to the knowledge and experience he gained from other manufacturers and mechanics within and outside the South, the third cornerstone of his southern manufacturing system. In fact, Merrell was instrumental in constructing pieces of those networks himself, partly because he was so confounded by the absence of people with mechanical skills in the South. His autobiographies showcase the series of networks this energetic man relied upon to make a success of cotton manufacturing. At the same time, we can see how Merrell himself contributed in important ways to the making of a southern technical community. While it is clear that Merrell believed his inherent mechanical ability was instrumental to his success, his willingness to embrace the latest technologies and especially his sustained knowledge as a machinist, a manufacturer, and an entrepreneur came from the circles of relationships he cultivated over the course of his career. Thus, his apprenticeship in Whiteside, New York, not only gave him a valuable base of knowledge about the cotton industry and manufacturing, it gave him access to similarly trained young men, whom early in his career he persuaded to join him in Georgia. "My own success depending upon a judicious choice of overseers, I resolved to have them from the North, and to have them selected among my own comrades, whose qualifications were known to me and I to them." Two of his Oneida friends joined him in 1839. Unfortunately one of them, a man named Parker, died from a fever acquired on the Georgia coast within days of his arrival. Merrell subsequently sought out other northerners with varying results. Still, Parker's unexpected death haunted him and helped persuade him to look for local men as well.[40]

40. Ibid., 165, 149, 165.

Merrell did not think most local mechanics, whether they were southern- or northern-born or trained, deserved inclusion in these networks. He based his opinion on their bad attitudes and poor skills. Barrington King had hired a number of mechanics to assist him with factory construction prior to Merrell's arrival. On meeting these hires, Merrell considered King "ignorant of mechanics" for he found he had accrued

> about him a bad-set kind from among the adventurers lying about the city of Savannah. They drew large pay &, in return, idled away a great deal of time, drank at their work, & were disorderly in their conduct. They were headed by a one-eyed sinner from N. York City—an Englishman named Hagg and well-named. He was a mill wright by trade. This gang were disposed to dictate unreasonable terms to their employers, & they were not at all disposed to accept me, a very young man, as their superintendent.[41]

Merrell resolved to fire the lot of them, especially when they kept slowing down as the work neared completion while insisting on more money. This put him in a difficult spot. Barrington King did not know whether to trust the mechanics' view of things or his new twenty-two-year-old assistant superintendent's. Merrell decided to establish his authority by setting up habits of industry governed by a clock. Mechanics who did not labor the full day as measured by an hourly factory bell would receive pay reductions in proportion to the time wasted. When Merrell held a powerfully built mill-wright named Mr. Atkinson to account for not working the requisite hours, he received a blow to the face that knocked him down a hill in front of several dozen employees. The two then took up fisticuffs, and after much exertion, Merrell won the fight. While the mechanics accepted his terms thereafter, Merrell remained skeptical of their work ethic and their skills. Having to "make out with mill wrights who denied the first principles of hydraulics and hydrostatics, or the strength of materials, with engineers who treated steam as a mystery" was not going to make his factory successful.[42]

Unimpressed with southern men who touted themselves as mechanics, Merrell elected to train his own, and he looked to planters' sons first. Although Merrell and his northern-born friends

41. Ibid., 151.
42. Ibid., 152–53, 170.

considered learning the mechanic trade to be a respectable, worthy career, especially for future manufacturers, he discovered elite and middle-class southerners did not. After finding only two sons of gentry planters willing to "turn their hands to mechanics," neither of whom would accept a seven-year apprenticeship nor his supervision, he lit upon his own operatives as mechanics-in-the-making. It took a great deal of time and patience for him to scout out suitable trainees, for he did not think highly of the average white male textile hand. He described his operatives in general, regardless of sex, as "the Gold-digging dirt-eating crackers of Georgia." He noted the adult men were often violent, carried knives and Colt pistols and proved "willing to use them at the drop of a hat." He did not think that those characteristics lent themselves to the acquisition of mechanical skills and managing abilities, but he was willing to, and did, train promising youths of southern birth who had grown up in the factories.[43] In this respect, Merrell had learned to operate within the realities of this planter-dominated society.

Merrell must have been a good teacher and a good employer to those who wanted to learn the trade. He had been disappointed by the poor quality of Georgia mechanics in 1839, but he was more forgiving of the Arkansas mountaineers he encountered in 1856, though he knew he would find no manager or mechanic he could employ among them. "Because they were so very poor and rude in their living, [they] possessed a certain rude skill in all the branches of industry necessary for their bare subsistence, nothing more." This time, seventeen years later, Merrell did not send to upstate New York or anywhere in the North for an experienced superintendent, but to Georgia. He remarked, "Such a man could . . . be found in Georgia or in the old [southern] states." He wound up hiring his last superintendent in Georgia, William Bell, noting, "Nothing could have been better timed or more suitable for me." Other mechanics formerly under him in Georgia also traveled of their own accord to Pike County, Arkansas, to work for Merrell, attesting to Merrell's important hand in developing a network of skilled managers and mechanics in the South.[44]

Merrell's account makes clear his presumption that most mechanical and all management work was for white men only. Black men,

43. Ibid., 166, 169, 170, 166.
44. Ibid., 259, 279–80.

slave or free, might learn adequate skills, but "however efficient, could not be placed in situations of responsibility, even among Negro hands." Nonetheless, he believed there was a place, if always a subordinate one, for skilled slaves and free blacks in factories. Therefore, Merrell was not averse to training African American men to be mechanics. "I never failed to make a good workman of a smart Negro, first having a care to select my man. Negroes have the manipulation for anything the most difficult." But he also presumed, like so many white men of his time and place, that talented slave mechanics must be made to understand their second-class status, even if that meant resorting to physical punishment.[45]

That the northern-born Merrell, who grew up in a section of the country that helped birth the antislavery movement, could accept the white southern outlook on slavery and the place of slaves in southern society with such apparent ease deserves comment. Another northern-born white man, Lewis Paine, whose circumstances and arrival in Georgia paralleled Merrell's own, had trained as a mechanic in cotton manufactories in Massachusetts and Rhode Island before moving to Upson County, Georgia, in 1841. Like Merrell, Paine had been contracted to start and run machinery in a factory—owned by D. R. Perry and Company in Paine's case. Upon his arrival, Paine was struck by the contrasts between northerners and southerners, just as Merrell had been. "Here all the modes of life, habits, and customs were marked by a striking difference from those of the North." Unlike Merrell, however, Paine observed that "the slaves excited my curiosity more powerfully than all other things." Also unlike Merrell, Lewis aided a runaway slave, albeit unintentionally, and for his efforts was imprisoned in Georgia for six years, during which he wrote an abolitionist tract based on his experiences, which was published in 1851.[46]

45. Ibid., 169–70. The contributions of slave inventors are just beginning to be recognized. See for example, William K. Scarborough's review of the exhibit "The Cotton Gin and Its Harvest," *Journal of American History* vol. 81, no. 3, The Practice of American History: A Special Issue (December 1994), 1238–43. Yet the WPA Narratives are full of examples like this one by former Arkansas slave J. H. Beckwith, "My father . . . had a mechanical talent. He seemed to be somewhat of a genius. He had a productive mind." Former slave Esther King Casey, who grew up in Americus, Georgia, recalled about her father: "Papa was a mechanic. He built houses and made tools and machinery." *American Slavery: A Composite Autobiography*, First Series, Library of Congress Rare Book Room Collection, Arkansas Narratives, Vol. 08A, 132; Alabama Narratives, Vol. 06A, 55.

46. Lewis W. Paine, *Six Years in a Georgia Prison: Narrative of Lewis W. Paine, Who*

That Merrell did not see the hypocrisy of extolling the virtues of free white labor in a society dominated by slave labor should not surprise us. The great majority of northern-born mechanics in antebellum Georgia did not help runaway slaves or advocate abolition, and they behaved cautiously when addressing the morality of slavery given native-born Georgians' suspicion of their motives.[47] Paine, not Merrell, was the exception. Merrell did acknowledge these blinders late in life. Commenting on why he had failed to recognize slaves as fully equal to whites, he remarked, "at that time my prejudices barred my way in that direction, & I will not say whether they were my own prejudices or those of others; perhaps we shared them all around."[48] Suffice it to conclude that Merrell did not consider African Americans key contributors to the networks of mechanics, managers, and manufacturers that he helped create and relied upon in the late antebellum South. Nonetheless, one cannot doubt African Americans' important service to the larger southern technical community, and certainly recent scholarship attests to their critical role. Merrell simply could not let himself see it.

Merrell's commitment to training black and white mechanics and white manufacturers-in-the-making was not unusual for this place or time. Samuel Griswold, one of the earliest cotton gin manufacturers in the deep South, who began his career as a tinsmith in Clinton, Georgia, had opened the first gin factory in the region by 1831. A northern transplant, Griswold trained scores of young men, many of whom later opened their own factories. The most celebrated of Griswold's trainees was Daniel Pratt, the New Hampshire-born carpenter's apprentice-turned-gin manufacturer-turned-Alabama industrialist and founder of Prattville, one of the most successful manufacturing communities in the South. Pratt employed several dozen mechanics, mostly southern-born, as likely to be slaves and free blacks as whites; his highest-paid mechanics were northern-born white men.[49] Transplanted northern mechanics and manufac-

Suffered Imprisonment Six Years in Georgia, for the Crime of Aiding the Escape of a Fellow-Man from That State after he had Escaped Slavery (New York: Printed for the author, 1851), 13–14.

47. Michele Gillespie, *Free Labor in an Unfree World, 1789–1860* (Athens: University of Georgia Press, 2000), chap. 5.

48. Skinner, ed., *Autobiography,* 166.

49. Gillespie, *Free Labor,* 108–9; Curtis J. Evans, *The Conquest of Labor: Daniel Pratt and Southern Industrialization* (Baton Rouge: Louisiana State University Press, 2001), 10–14.

turers played a critical role in introducing mechanical knowledge and engineering principles to the broader community. We are only beginning to recognize what that transformation of knowledge meant for the growth of a southern middle class over the course of the nineteenth century.

The final cornerstone of Merrell's system centered on his insistence that manufacturers and capitalists share knowledge with each other. Southern manufactures were competing in an ever-widening market, and this reality necessitated better cooperation with one another. In fact, Merrell was surprised that southern manufacturers did not band together more. He felt southerners in general were poorly treated by Yankees, who often sold them poorly made goods, from shoes to machine works. He believed northerners did so because southerners advocated slavery.[50] Although capitalistic in many respects, even as they remained paternalistic and patriarchal in others, southerners were often blind to the value of cooperating economically with one another at the local level. Merrell found this degree of individualism frustrating.

While Merrell sustained most of his networks through his personal relationships (family, church, friends, employees, suppliers, acquaintances, etc.), he resorted to print culture to advocate his southern system too, because he believed southern manufacturing would only improve with more open exchanges of information. "I wrote articles for the public prints over a non de plume, & I was an attentive correspondent with all who wrote me on the subject of Manufacturing, imparting freely, without remuneration, information it had cost me time and money to obtain. In all this I am quite sure I had no aspirations except to be useful."[51]

In his 1847 series for the *Southern Recorder* titled "Manufacturing in Sober Earnest," Merrell exhorted Georgia manufacturers to pay closer attention not only to the market but to one another. Too many start-up factories were making thread, for example, resulting in overproduction that drove down everyone's profits. He understood the impulse to strike out on one's own, heedless of the competition, since manufacturing was in its infancy in the region. He acknowledged that Georgia had a long way to go compared to England and the North. But he also reported that people in the countryside were

50. Skinner, ed., *Autobiography*, 188.
51. Ibid., 167.

now accruing more wealth, which gave them more buying power and made them less willing to rely on home production. The wisest manufacturers "will produce better goods at less price, and even take produce in return. Such has been the invariable result in other sections of our country where manufacturing and the mechanic have the time and opportunity for development." Manufacturers, he said, must be clever given these promising developments and avoid driving themselves into bankruptcy.[52] Moreover, and ironically, Merrell contended, "We are now suffering from over production and the other evils of high competition, not less than the old established manufactories of the Northern States." The largest factories with the most capital to finance the most improved machinery would always succeed in any region of the country. To be successful in this climate of changing cotton prices required diversification of products and flexibility. Manufacturers needed to drop those products that generated no profits and respond to the latest demands of the market.[53]

But even more than diversification and flexibility, Merrell advised, southern manufacturers had to acknowledge that their "greatest difficulty in manufacturing grows out of our relations toward each other." If manufacturers would work together for their "mutual aid and comfort," they would discover access to substantial and hitherto untapped resources. By exchanging aid, reciprocating favors, rendering prompt assistance in times of accident or danger, sharing facilities for repair, even simply engaging in conversations "with men superior to yourself in science, skill and capacity, with whom you consult about your difficulties and debate the merits of improvement," and by establishing more uniform wages and disciplining of hands, the general good would be more fully served, even as individual manufacturers would benefit greatly through this cooperation. "Something of a community must be made of the manufacturing interest here, in order to have any further advancement here or even to hold that which we have already achieved," he advised. And while it made sense to look to the northern example, he believed that southern manufacturers had to appreciate the uniqueness of southern manufactories and use their own wisdom

52. "Manufacturing in Sober Earnest: No. 1," *Southern Recorder (Milledgeville)*, March 30, 1847.
53. "Manufacturing in Sober Earnest: No. 2," *Southern Recorder (Milledgeville)*, April 13, 1847.

and experience to make the right decisions.[54] Merrell, of course, acted on his own advice.

Eight years later, Merrell wrote another essay series on southern manufacturing for the *New York Journal of Commerce*. He bragged about cotton manufacturing's success in Georgia and showed how planters and the state political economy benefited from industry. While he admitted that some of the current fifty or so factories were badly managed, badly constructed, and badly capitalized, many others boasted the best machinery and construction found anywhere in the nation. The capital for these efforts had come from cotton profits, diverting capital (as did railroads) from more cotton planting through traditional investments in land and slaves, and thereby reducing the cotton supply and driving up cotton prices on the open market. Thus, Merrell pointed out, southern manufacturing was valuable to the South as a whole and supported planters' interests in the end.

He also argued that factories "have forced into active employment, and into something like discipline, a very unruly and unproductive class of white population, who, when idle, are, to say the very least, no friends of the planter. There are now no paupers, to speak of, in any county of Georgia where a factory exists." Thus, the state of Georgia could avoid instituting a poor tax thanks to the gainful employment of the poor in factories. While this conclusion was not exactly accurate, Merrell certainly recognized that his audience would be receptive to this alleged benefit to the state economy.[55] Merrell did not suffer fools lightly before this national audience. He also mocked southern capitalists who entered the field as a lark:

> People go into it half cocked, not investing enough, choosing bad locations, and yet a little money is made nonetheless—is really a curiosity and plaything for the stockholders. They cannot keep away from it [the dabblers]. They purchase books on the subject, and read them. They bring troops of friends, unto whom they explain the mechanical operations, not perhaps in the most lucid manner, yet with sufficient unction . . . and [they] get outraged when they cannot smoke their cigars in the lint room—when dividends returned

54. "Manufacturing in Sober Earnest: No. 3," *Southern Recorder (Milledgeville)*, April 20, 1847.

55. "Cotton Manufacturing in Georgia: No. 1," *New York Journal of Commerce*, August 31, 1855, p. 3: col. 3.

diminish, after first being high and encouraging others to go into the business, the manager chased out (always a Northerner) and the business falls apart.[56]

Merrell was accurate in his assessment of the commitment of many planter-manufacturers. The Vaucluse factory in the Edgefield district of South Carolina for example, suffered serious financial setbacks in the 1830s that can only be attributed to the failure of investors to commit themselves permanently to cotton manufacture.[57]

Merrell was casting aspersions on planters, but he was also implicitly extolling the value of seasoned, knowledgeable mechanics and managers in leadership positions. He cautioned that the success of the largest factories in Georgia and South Carolina was due to this class of men, who invariably could be recognized by their prematurely gray hair and deeply lined faces, signs of their responsibility and wisdom. All manufacturing failures in the South, he concluded, could be chalked up to "amateur manufacturing by gentlemen who know nothing about the business." The future of southern manufacturing lay in recognizing the centrality of that professional class of manufacturing men and their cooperative networks.[58]

Not all southern industrialists agreed with Merrell's conclusions, which involved a degree of abstraction not necessarily embraced by more practical men. Daniel Pratt, one of the South's most famous manufacturers, was less concerned with building technical communities than with competing successfully with northern textiles. When Alabama merchants complained about the high prices of southern-produced textiles, Pratt reminded them that southern factories had to purchase northern machinery, pay for the shipping and freight, and hire "Yankee superintendents and machinists," who were not cheap, all driving up the cost of goods.[59] He had no intention of cooperating with his less successful neighbors against northern interests.

---

56. "Cotton Manufacturing in Georgia: No. 2," *New York Journal of Commerce,* September 6, 1855, p. 1: col. 1.

57. Tom Downey, *Planting a Capitalist South: Masters, Merchants, and Manufacturers, 1790–1860* (Baton Rouge: Louisiana State University Press, 2006), 126–28.

58. "Cotton Manufacturing in Georgia," nos. 2, 3, and 4, "Cotton *New York Journal of Commerce,* September 6, 1855, p. 1: col. 1; September 8, 1855, p. 3: cols. 4–5; September 17, 1855, p. 3: col. 3.

59. Evans, *Conquest of Labor,* 71–72.

By contrast, Merrell styled himself a simple self-made man of faith imbued with expert mechanical know-how, and hence he eschewed many of the traditions he encountered among the southern aristocracy: "the elaborate manners of southern gentlemen seemed very stupid to me." On the other hand, he was committed to securing relationships that created technical knowledge for himself, which he willingly shared with other southern manufacturers. He was not opposed to some self-aggrandizement, which he glazed with a touch of self-pity for pursuing this altruistic role. "Truly he who sets out to be a leader in new enterprises running counter to the genius, and cutting across the prejudices of a people—and that twenty years ahead of his time—has an arduous life!"

Merrell's expertise with machinery and his connections with mechanics and manufacturers around the country proved integral to his success and were valuable commodities in this southern world. These attributes, along with his fierce work ethic, led him to incorporate the latest technological changes in two of his four textile factories, assuring him profits in an increasingly mercurial and competitive market. He understood that his professional accomplishments rested in large measure on his impressive network of technological knowledge. His business crises, the bankrupted Greene County textile mills, resulted not from the failure of those networks of knowledge, but from his decision to ignore them, which he later humbly acknowledged.

> Although I possessed a decided genius for inventions, & had the peculiar gift of elaborating an improvement in my brain, and bringing it to perfection without much experiment, I went so far as to conceive that I had an aversion to improvement in general, which finally led me to one or two memorable failures in the State of Georgia, by using old-time machinery at a conjuncture when I should have adopted every genuine improvement, and taken all the chances for success. Hence it was that in Georgia, my own pupils, in their turn, profiting from my errors, out-ranked me in my own profession, and made their fortunes while I was losing mine. I learned wisdom at last, and in Arkansas, I have tried to hit the medium proper, and I have had success.[60]

60. Skinner, ed., *Autobiography*, 243.

Merrell proved an important node in this developing southern technology network. He not only built and managed three textile mills in Georgia and a fourth in Arkansas, but he also committed himself to thoroughly training his assistants. Not only did they learn how to be skilled mechanics and talented managers, but Merrell tutored them on the principles of his southern manufacturing system as well. As Merrell points out above, these young men—"my own pupils"—used that knowledge to beat him at his own game in the late 1840s. This irony notwithstanding, Merrell made a significant contribution to the development and spread of a southern technical community through his commitment to the next generation of manufacturing men.

Richard D. Brown has argued that a communications revolution—the dissemination of information through printed material, oratory, steamboat, railroad, and telegraph—sharply accelerated the pace at which information could travel, as well as the democratization of information, taking it out of the controlling pockets of the wealthy, learned, and powerful and putting it into the hands of everyday people. By the Civil War era, Brown concludes, Americans had constructed, with the help of new technology and an expanding economy, a "pluralistic information marketplace."[61] Henry Merrell understood that his knowledge as a machinist and a manufacturer, and his insistence on keeping up with and utilizing changing technologies in his business—knowledge secured and updated through his networks of relationships—gave him a kind of limited authority in the agricultural slaveholding South.

There can be no question that the South was hampered by its relative inexperience at machine making during the antebellum era. The greatest innovations in textile machinery in the United States and in Britain took place in the two decades before the Civil War. These innovations prompted the separation of textile manufacture and textile machinery. Improvements no longer depended upon the efforts of textile owners and managers themselves, but upon the inventiveness of northern machine works in industrial towns, especially Paterson, but also Philadelphia, Taunton, Providence,

---

61. Richard Brown, *Knowledge Is Power: The Diffusion of Information in Early America, 1700–1865* (New York: Oxford University Press, 1989). The value of Brown's argument lies in its insistence on the spread of many different kinds of discourse. Brown's book says virtually nothing about the antebellum South per se and very little about the new class of men like Henry Merrell.

Worcester, and Lowell. The South had begun to produce fledgling industrial towns, such as Richmond and Macon, but the northern places benefited from the sheer number of companies, experienced and new, the availability and relative cheapness of artisans and mechanics, and the emergence of engineering as a profession.[62] The South was not there yet, but men like Merrell were laying the groundwork for postbellum development.

Henry Merrell's desire to construct a southern-based network of knowledge and to contribute to the building of a southern technical community inherently challenged a hierarchical society where property, especially in the form of land and slaves, had long been the best assurance of authority. The successes that Merrell accrued from putting his ideas into practice during his three decades as a cotton superintendent and manufacturer in the antebellum South and his generosity in training the next generation of mechanics and manufacturers helped transform technological innovation into an increasingly valued concept in the late antebellum South. Indeed, we need to recognize that the efforts of men like Henry Merrell help explain why the economic interests of merchants, manufacturers, and capitalists began to hold sway over those of agriculturalists and planters on the eve of the Civil War.[63]

62. Victor S. Clark, *History of Manufactures in the United States,* vol. 1, 1607–1860 (Washington, DC: Carnegie Institute of Washington, 1929, reprint, New York: Peter Smith, 1949), 434–35.
63. Tom Downey makes this argument in *Planting a Capitalist South.*

Entrepreneurial Networks and the Textile Industry

Technology, Innovation, and Labor in the American Southeast, 1890–1925

PAMELA C. EDWARDS

SOME HISTORIANS, IN ANALYZING THE NATURE AND DIFFUSION of industrialization, have stressed its fluid and regional nature. They describe a regional ebb and flow of development that caused some locales to briefly flourish only to be displaced when capital, labor, and/or technological advantage shifted to a new region of the state or globe. These historians do not describe industrialization as a single trajectory. Instead, they document a messier business of overlapping and very human interests, where capital interacts with invention, technological diffusion responds to labor relations, where politicians and industrialists shake hands and serve on the same boards or committees. One generation initiates industrial empire, building a network of transportation, banking, factory, and

A substantial portion of the research presented in this essay has previously appeared in *South Carolina Historical Magazine* and *North Carolina Historical Review*. This article, however, consolidates and combines the information dealing with Columbia and Lyman, South Carolina, with material dealing with the hosiery industry in Carrboro, North Carolina, to more carefully consider the behavior and actions taken by entrepreneurs in different southeastern communities. While emphasizing diverse community and entrepreneurial responses, the actions of unions, workers, and those who would hinder their organization in these communities receive more attention than in the earlier articles. Please see Pamela C. Edwards, "'In Good Faith': The Rise and Fall of a Company Union in the Durham Hosiery Mills," *North Carolina Historical Review* vol. 81, no. 4 (October 2004): 365–92 and Pamela C. Edwards, "Southern Industrialization and Northern Industrial Networks: The New South Textile Industry in Columbia and Lyman, South Carolina," *South Carolina Historical Magazine* vol. 105, no. 4 (October 2004): 282–305.

machinery manufactures essential for industrial growth that inevitably matures and then slips into decline. Regional cores of early development give way to competition from underdeveloped regions when their maturity advances production costs to a point where less developed regions can take advantage of lower labor costs and access to raw materials.[1] The closer historians get to the internal dynamic of regional industrial development and decline, the closer they get to the human interactions taking place when an underdeveloped region breaks into an industry, displacing the original center of industrial revolution in that industry, the closer they get to understanding the dynamic of capitalist development and how underdevelopment may, perhaps, be alleviated and the labor and environmental exploitation and damage so often associated with development controlled or prevented.

In the first volume of this series, *Global Perspectives on Industrial Transformation in the American South*, historians David Carlton, Peter Coclanis, Beth English, and others apply this approach to the study of industrial development to the American South. In "Southern Textiles in Global Context," Carlton and Coclanis, building on the work of Gavin Wright, stress the global character of the southern economy and place the development of the southern textile industry within its international context. At the same time, they compare the regional development of the southern textile industry with the simultaneous nineteenth- and early-twentieth-century development of the textile industry in Brazil, India, and Japan.[2] Similarly, Beth English, in "Beginnings of the Global Economy: Capital Mobility and the 1890s U.S. Textile Industry," examines the process by which the Dwight Manufacturing Company, a New England–based textile firm, moved its manufacturing facilities to Alabama, noting the efforts of Alabama businessmen, probusiness politicians, and community promoters to lure the northern textile firm to Alabama City with tax exemptions, low-wage, nonunion labor, and

1. Maxine Berg, *The Age of Manufactures 1700–1820: Industry, Innovation and Work in Britain* (London: Routledge, 1994); Sidney Pollard, *Peaceful Conquest* (Oxford: Oxford University Press, 1981).

2. David L. Carlton and Peter Coclanis, "Southern Textiles in Global Context," in *Global Perspectives on Industrial Transformation in the American South*, ed. Susanna Delfino and Michele Gillespie (Columbia: University of Missouri Press, 2005), 151–74. See also Gavin Wright, *Old South, New South: Revolutions in the Southern Economy Since the Civil War* (New York: Basic Books, 1986), 124–25, 269–74.

state legislative resistance to protective labor laws.[3] Historians of industrialization give much attention to the textile industry because textiles led the first industrial revolution in England and the United States. The textile industry was the catalyst industry for development in many countries and regions around the globe, including the American South.

In keeping with the focus on southern entrepreneurs established for this volume, this essay examines three entrepreneurial efforts made by business and community leaders in southeastern communities between 1890 and 1925. In all three situations, entrepreneurial efforts succeeded in establishing textile manufacturing facilities, bringing jobs and new businesses to their communities. In Columbia, South Carolina, engineer and entrepreneur W. B. Smith Whaley established four state-of-the-art textile factories by making use of his ties with northern textile engineering and business leaders. A second approach was taken by entrepreneurial members of the Spartanburg, South Carolina, Chamber of Commerce, who lured a large northern-based textile firm to their community with tax incentives and promotion of their natural resources and abundant supply of low-wage labor. The third case study considers the entrepreneurial efforts of Julian Carr Sr. and his sons in Carrboro, North Carolina, where their focus on the niche industry of hosiery manufacture and development of regional business and technological networks allowed the Durham Hosiery Mills to compete successfully against long-established northern manufacturers. In all three situations, southern entrepreneurs balanced regional advantages of abundant and cheap supplies of labor and land against disadvantages in available capital, technical skills, available transportation, and marketing agencies. Cutting across these case studies are two additional phenomena of southern entrepreneurship, one dealing with the relocation of an industry and the other with southern labor networks. First, the entrepreneurs all had to contend with the New England–based Pacific Mills Corporation, which eventually arrived in all three of their communities. Initially located in Lawrence,

---

3. Beth English, "Beginnings of the Global Economy: Capital Mobility and the 1890s U.S. Textile Industry," in *Global Perspectives on Industrial Transformation in the American South*, ed. Susanna Delfino and Michele Gillespie (Columbia: University of Missouri Press, 2005), 175–98. In *A Common Thread: Labor, Politics, and Capital Mobility in the Textile Industry* (Athens: University of Georgia Press, 2006), English expands on the ideas in her earlier essay.

Massachusetts, Pacific Mills relocated its corporate assets to the American Southeast between 1916 and 1950. As the company opened southern branches, it slowly closed all its northern facilities and, in doing so, illustrated the regional relocation of an industry. Second, these three entrepreneurial efforts indicate two networks linked to southern labor—one made up of businessmen and politicians who sought to control the southern industrial labor force and another rising from labor and representing workers' efforts to unite with national and international labor organizations in opposition to capital. An examination of these three case studies reveals multiple paths to community industrial development in the relatively underdeveloped southeast between 1890 and 1925. We discover a complex array of entrepreneurial endeavors and multiple paths to community industrial development, each path requiring local compromise with established northern financial, technological, and business networks of the mature northern-based textile industry.

W. B. Smith Whaley in Columbia:
Southern Industry, Northern Networks

W. B. Smith Whaley was born in 1866 and grew up in Charleston, South Carolina, but he received his higher education in New York and New Jersey. He attended Bingham Military Institute (New York), the Stevens Institute of Technology (New Jersey), and Cornell University. From Cornell he received a degree in mechanical engineering in 1888. After graduating, he worked for the Daft Electric Company and, later, for Thompson and Nagle Engineers, of Providence, Rhode Island. While employed by Thompson and Nagle, Whaley helped to develop the water supply system in Columbia, South Carolina. The potential for the production of electricity in the Columbia area attracted manufacturers to the city in the 1890s, including Whaley's textile plants. After leaving Rhode Island, Whaley returned to his home state to work for a phosphate company in Charleston, and in 1892, he opened W. B. Smith Whaley and Company, a cotton mill engineering firm.[4] The firm's success and

---

4. Fenelon DeVere Smith, "The Economic Development of the Textile Industry in the Columbia, South Carolina, Area from 1790 through 1916" (Ph.D. diss.,

expansion relied in part on Whaley's ongoing personal contact with northern engineers with whom he had gone to school and worked in New England. He would eventually use these contacts to initiate his plans for textile manufacturing in Columbia. Between 1895 and 1904, Whaley founded, built, and placed in operation four cotton textile factories—the Richland, Granby, Capital City, and Olympia—in Columbia, South Carolina.[5]

Constructed in the 1890s, the Whaley mills were carefully located to take advantage of regional technological and capital assets. In terms of technology, Whaley hoped to utilize power and transportation resources and emulated the pattern of industrial development followed by the New England textile industry. Using his connections with northern engineering firms, Whaley simultaneously sought to establish local technological independence while simultaneously ensuring access to northern technological networks. Though he established local transportation and power sources, he also relied heavily on northern machinery manufacturers and technical expertise. He arranged loans to bring his Columbia mills into production, working initially with members of his board of directors, which in the beginning was primarily made up of Columbia-based businessmen.[6] As time passed, however, Whaley was forced to abandon his hopes for maintaining local control over the technological and capital assets of the Whaley mills in Columbia. To acquire the technology needed to operate the four mills, he traded stock subscriptions for supplies, including building materials, textile machinery, and raw cotton. In so doing, he extended ownership in the Columbia companies to machinery manufacturers in the Northeast and to cotton factors based in New York. When he refinanced company debts, he did

---

University of Kentucky, 1952), 116–20. The history of the textile industry in Columbia, South Carolina, discussed in this article draws heavily on Smith's dissertation. Drawing upon Smith's history of the development of the Whaley mills and Pacific Mills' eventual takeover of them, this essay places those events within the context of national and regional economic networks. See also John Hammond Moore, *Columbia and Richland County: A South Carolina Community, 1740–1990* (Columbia: University of South Carolina Press, 1993), 305-6. For an overview of industrial development in South Carolina during the New South era, see David L. Carlton, *Mill and Town in South Carolina, 1880–1920* (Baton Rouge: Louisiana State University Press, 1982).

5. Smith, "Economic Development," 118–79. See also Alvin W. Byars, *Olympia-Pacific: The Way It Was, 1895–1970* (Columbia, SC: Professional Printers, Ltd., 1981), 1–4.

6. Smith, "Economic Development," 120–203.

so with loans and bond issues administered through New England banks. These financial institutions were owned and operated by the same men who ran the textile factories in Lowell, Lawrence, and other New England industrial cities. In the end, his increasing reliance on northern financial and technological networks resulted in, essentially, handing over controlling interest in his firms to members of the institutions that controlled those networks.[7] Whaley's failure to establish independent regional capital and technological networks required the Columbia business community to seek an alternative path toward industrialization, one characterized by northern ownership and management of southern firms. Intricate ties developed between leaders in New England textile and financial circles and the textile industrialists of Columbia. An exploration of these personal, financial, and technological ties provides insight into the relationship between New South industrial development and northern financial, marketing, and technological networks.

Northern engineering strongly influenced Whaley's education and career. It is not surprising that the mills he constructed in Columbia were based on a long-established New England pattern.[8] Both architecturally beautiful and functionally innovative, the four Whaley mills were superbly crafted with the best materials, so much so that, when the mills faced financial crises, some accused Whaley of going to unnecessary expense in their construction.[9] One description of the construction techniques used in the Olympia, the largest and most experimental of the factories, suggests the sophistication of Whaley's design:

> The walls and towers of the mill were built of red brick laid with red mortar. The first floor of the mill was constructed of concrete and board with a finished floor of maple. Steel beams and concrete arches were used to construct the floor over the dust room chambers, wire tunnels and first floor of the rear towers. Except in the rear towers, the basic construction was covered with a maple finishing floor. The

7. Ibid. See also Carlton and Coclanis, "Southern Textiles in Global Context," 154-64; and Wright, *Old South, New South,* 125, 131. Both discuss the position of U.S. textile machinery manufacturers and why they invested, via stock subscriptions, in southern textile companies in the late nineteenth and twentieth centuries.

8. Smith, "Economic Development," 116–20.

9. Anonymous, Unbound Scrapbook of Reorganization, 1903–1905, Manuscript Collection, South Caroliniana Library, University of South Carolina, Columbia, South Carolina, hereinafter SCL.

floors in the rear towers were covered with Venetian mosaic. For the construction of the second, third, and fourth stories, 3 3/4 inch spliced pine were used, covered by a 7/8 inch diagonally-laid intermediate floor, and topped with a 7/8 inch maple floor, the layers of flooring being interspersed with tarpaper. The mill and power plant roofs were covered with asphalt gravel roofing.[10]

In his Columbia mills, Whaley included recent techniques in textile factory design, such as concrete and steel construction, a system of steam coils to regulate heat, a fire prevention sprinkler system supplied by two fifteen thousand–gallon tanks and a humidification system located on top of the mill, which supplied moistened air to all four floors.[11] In addition, Whaley attempted innovations of his own. Deciding that the Columbia Water Power Company could not provide sufficient power at a reasonable rate, Whaley designed Olympia Mills "for an electric drive, with electric generators directly connected to shafts of steam engines, and with electric motors suspended from the ceilings."[12] He located a large power plant at the rear center of Olympia Mills and used the electric power generated there to run the machinery and furnish lights to all four Whaley mills. To supplement the power of the Granby and Richland mills and to fully supply the Capital City Mills and the Columbia Street Railway Power Company's circuits, Whaley installed a circuit switchboard, making Olympia Mills the first textile mill in the United States to be fully powered by electricity.[13] Not all of Whaley's experiments, however, paid off, and as late as 1905, the company president reported to stockholders that the "power plant . . . has continued to cause problems," leading to shutdowns, decreased production, and increased costs. In 1906, the company spent over three thousand dollars in an attempt to repair the Olympia power plant.[14] In a letter to stockholders dated January 5,

10. Byars, *Olympia-Pacific*, 12.

11. Ibid.; Smith, "Economic Development," 153–77; *Modern Cotton Mill Engineering* (Columbia, SC: State Company, 1903), 31–47.

12. Byars, *Olympia-Pacific*, 12; Smith, "Economic Development," 166–75.

13. Smith, "Economic Development," 166–75; Byars, *Olympia-Pacific*, 12–15; Jacquelyn Dowd Hall et al., *Like a Family: The Making of a Southern Cotton Mill World* (New York: W. W. Norton & Company, 1987), 48; *Modern Cotton Mill Engineering*, 31–47.

14. President's Report to Stockholders, November 20, 1905, Olympia Cotton Mills, Minute Book of Stockholders' Meetings, 1899–1912, 83–89, SCL; President's Annual Report to Stockholders, November 15, 1906, Olympia Cotton Mills, Minute Book of Stockholders' Meetings, 1899–1912, 90–99, SCL.

1906, president Lewis W. Parker, Whaley's successor, reported: "[a] difficult [year] on operations, owing to the violent fluctuations in cotton and great changes in its value, and also. . . to the continued inconstancy of our power. . . negotiations are now in progress which will insure . . . provisions for additional power, which will prevent the interruptions referred to."[15]

The tension between dependence on northern machinery manufacturers and efforts to maintain independent local control over southern industry can be observed through machinery purchases and installation. Once the factory buildings were constructed, Whaley faced the problem of purchasing, receiving, installing, and maintaining production machinery. Whaley equipped his Columbia textile mills with the most advanced production machinery available in the 1890s; all were constructed in New England and shipped to Columbia.[16] While Whaley had connections in New England engineering and textile manufacturing circles, his connections were strained by distance. Transportation between the Northeast and the South was slow and it was difficult to gage when machinery orders would arrive. In addition, if machinery arrived in need of adjustment or repair, servicing required either returning it to the manufacturer or arranging for a service representative to make an on-site visit. Either scenario could delay installation by weeks or months and decrease production and profit margins.

Transportation expenses plagued the first two presidents of the Whaley mills and made it difficult to acquire the manufacturing equipment needed to maintain efficient production levels. When the Granby Cotton Mill began operations, only part of the necessary machinery had been installed, and only surplus electrical power from Olympia Mills was available to power the new factory.[17] Even after equipment arrived and was installed and operating in one section of the mill, the new installation often offset the coordination of production levels from one department to the other, particularly from spinning to weaving.[18] In 1908, Parker reported the installa-

---

15. Parker to the Stockholders of the Granby Cotton Mills, January 5, 1906, William Guion Childs (1850–1912) Papers, 1848–1912, SCL.

16. Smith, "Economic Development," 128–29, 132–33, 143–45, 176–79. See also Carlton and Coclanis, "Southern Textiles in Global Context," 154–64; and Wright, *Old South, New South*, 125, 131, 144.

17. Byars, *Olympia-Pacific*, 5, 9.

18. For a discussion of this common problem, see Nancy Frances Kane, *Textiles*

tion of fifty-eight additional looms in Olympia Mills. He recommended to the board of directors that the company purchase one hundred more, explaining that the spinning and carding departments produced more yarn than could be woven "with the looms now at our disposal."[19] The difficulty and expense of transporting machinery from New England to Columbia and finished products from Columbia to New York selling agencies led Parker to invest in a boat line to New York, which he thought would provide cheaper transportation rates than the railroad. Parker explained to the board of directors that the "prevailing rate via railway is 41 cents per cwt [cotton weight ton], but confidentially, I think I can arrange to transport our freight by the boat line to New York for about 34 cents per cwt." Parker loaned the boating concern five thousand dollars, which was to be repaid gradually through shipping charges. Even in this case, the boat had to be loaded in Columbia and its cargo shipped to Georgetown, where it was reloaded onto ships of the Clyde Line for the trip to New York.[20]

In addition to transportation problems, the lack of adequate capital exacerbated the Whaley mills' burden in dealing with northern machinery manufacturers. In his third annual report to stockholders, Whaley expressed his frustration at trying to bring a mammoth enterprise like Olympia Mills into production without the necessary capital: "In the past year we have succeeded in very nearly filling our mill with machinery and getting practically the whole mill in operation . . . attaining . . . the large number of people necessary to furnish the help for an institution of the dimensions of the Olympia is no easy matter [and the problems of] bringing the mill to a uniform producing plant are legion."[21] Whaley sought to circumvent cash-flow difficulties by encouraging textile machinery manufacturers to accept the preferred stock of his Columbia factories in part payment for machinery. In 1907, four of nine owners of Olympia Cotton Mills common stock were machinery manufactur-

in Transition: Technology, Wages, and Industry Relocation in the U.S. Textile Industry, 1880–1930 (New York: Greenwood Press, 1988).

19. President's Annual Report to Stockholders, November 17, 1908, Olympia Cotton Mills, Board of Directors Meeting Minutes, 1899–1912, 100–103, SCL.

20. Parker to Executive Committee, September 27, 1906, William Guion Childs Papers, SCL.

21. President's Annual Report to the Stockholders, February 13, 1902, Olympia Cotton Mills, Minute Book of Stockholders' Meetings, 1899–1912, 26, SCL.

ers. At the same time, ten of forty-eight holders of second preferred stock can be easily identified as machinery manufacturers, and together they owned almost two-thirds of 6,790 shares. Banks or machinery manufacturers held the most first preferred Olympia stock. The Draper Corporation owned 1,230 shares of first preferred stock, Fales and Jenks Machine Company controlled at least 1,300 shares, and the Eastern Machinery Company owned 303 shares. All of these shares were exchanged for machinery and other equipment needed to bring the mill into full production.[22]

Between 1890 and 1916, Whaley, his associates, and Parker faced serious problems with transportation, communication, and production techniques. At various times, both stockholders and the media blamed the problems of the Whaley mills on poor management, and both Whaley and Parker were accused of mishandling stockholder investments. In most cases, the actions identified as mishandling were the men's attempts to circumvent capital shortages and technological problems. At the time of the first reorganization, one local newspaper accused "the directors and officials of Columbia's cotton mills" of "chenanigins [*sic*]" and salting "down a considerable pile for a rainy day."[23] Others blamed the "Olympia downfall" on the fact that it was "managed and controlled by an engineer and not by a manufacturer."[24] According to critics, the Whaley mills' directors had attempted to "build and equip [the] largest cotton mill in the world," Olympia Mills, on credit and at the expense of stockholders in the Granby and Richmond enterprises. Among the liabilities listed for Olympia Mills were "machinery builders who sold machinery part on cash, part on note under false pretense of directors that [the] company was in good shape."[25]

When Whaley returned to South Carolina and began overseeing the planning and construction of mills in Columbia, he sought to

22. Olympia Mills, List of Stockholders, February 27, 1907, Whaley Mills Reorganization Papers, 1907, SCL; Parker to Elliott, n.d., William Elliott Jr. (1872–1943) Papers, 1903–1907, SCL.

23. "The Whaley Mills," 1903 newspaper article, anonymous, Unbound Scrapbook of Reorganization, 1903–1905, SCL. This scrapbook includes newspaper articles concerning reorganization of the Whaley mills and the lawsuits initiated because of that reorganization.

24. "The Cotton Mills of the Piedmont," *Greenville News*, n.d., anonymous, Unbound Scrapbook of Reorganization, 1903–1905, 2.

25. "A Restraining Order Asked, Complaints in the Famous Olympia Mills Case," and "Whaley Mills Reorganize: Olympia, Richland and Granby Decide What to Do," newspaper articles, n.d., ibid., 4–10, 19–23.

recreate the technological networks that supported New England manufacturing centers, including power supplies and access to railroad systems. Whaley first considered Columbia a feasible southern industrial site because of the existing water-power system he had helped to engineer for the city. By the time he began to design textile factories in Columbia, he sought to incorporate as many aspects of a secure manufacturing network as possible. He located the mills near the intersection of several railroads that served the Columbia area, arranged to provide electric power to all four facilities, and constructed a state-of-the-art mill village to attract southern labor.[26] But Whaley's position was much more isolated than that of New England textile and machinery manufacturers. He depended upon the success of the Olympia powerhouse, and when it failed to meet his original vision, production and finances suffered at all four facilities. Unlike the Boston Associates, Whaley did not own or help to manage the railroads upon which he relied to deliver machinery and raw materials and to ship out his finished product. The distances between his mills in Columbia and machinery manufacturers and selling agencies was much greater than those between nineteenth-century New England textile factories and the port cities of Boston and New York. These disadvantages helped to undermine the financial stability of the Whaley mills, which led, in turn, to the loss of local control and a major New England textile firm's purchase of the Whaley mills in 1916.

The circumstances leading to Pacific Mills' takeover of the mills in Columbia are made clearer when the financial arrangements surrounding the construction and operation of the four Whaley mills are considered. The four mills dwarfed all other factories in the Columbia area in terms of size and design. Constructed one immediately after the other, each mill suffered from severe capital shortages and, as each project multiplied obligations, the financial condition of the entire enterprise grew increasingly precarious. All four firms suffered from a lack of fluid capital, and efforts to remedy this situation led to increasingly desperate measures. Attempts to overcome recurring financial crises tended to create interlocking boards of directors and a risky mixture of the personal finances of individual investors with the corporate finances of the textile

26. Smith, "Economic Development," 116–93. See also Robert F. Dalzell Jr., *Enterprising Elite: The Boston Associates and the World They Made* (Cambridge: Harvard University Press, 1987).

companies.[27] The firms' original stock remained unsubscribed, and the companies lacked adequate capital to complete construction and furnish the factories once they were built. Whaley and the directors of each firm endorsed personal notes in exchange for remaining stock, securing their consequent personal debt obligations by mortgaging mill properties.[28] When these loans proved inadequate, Whaley and his associates increased the capital stock of each company and used the new stock to purchase machinery and supplies. No one ever paid cash for these stock subscriptions. Instead, Whaley used the stock to pay machinery manufacturers for spinning equipment, looms, flooring, windows, and all sorts of mill furnishings.[29]

Despite these elaborate efforts, the Olympia, Capital City, Granby, and Richland mills continued to experience difficulties,

27. Richland Cotton Mills, Application for Charter, January 17, 1895, Records of the Secretary of State, Corporate Charter Division, File #975, Charter Book F, 16; Articles of Incorporation, February 12, 1895, Charter Book D, 578; Certificate of Organization, February 12, 1895, Charter Book F, 26; Richland Cotton Mills, Application for Increase of Capital Stock to $300,000, December 19, 1899, File #975, Charter Book P, 1–3, and Charter Book H, 96; Granby Cotton Mills, Petition for Charter, May 30, 1895, File #1038, Charter Book F, 80; Articles of Incorporation, September 11, 1895, Charter Book D, 648; Certificate of Organization, September 11, 1895, Charter Book F, 145, and Charter Book D, 648; Petition for Amendment of Charter for Increase of Stock, August 22, 1896, File #1038, Charter Book F, 317; Petition for Amendment of Charter for Increase of Stock, August 31, 1898, Charter Book H, 45, and Charter Book L, 105; Olympia Cotton Mills, Application for Charter, May 16, 1899, File #1717, Charter Book L, 281; Return of [In]corporators, July 3, 1900; Articles of Incorporation, August 3, 1899, Charter Book G, 455; Amendment of Charter for Increase of Capital, August 2, 1900, Charter Book H, 132, all in the South Carolina Department of Archives and History, Columbia, South Carolina, hereinafter SCDAH. For a list of the original incorporators, see August 1, 1899, August 4, 1899, June 20, 1900, Olympia Cotton Mills, Minute Book of the Stockholders' Meetings, 1899–1912, SCL. See also Smith, "Economic Development," 120–59; and Byars, *Olympia-Pacific*, 5–11. Constructing Richland Mills first, Whaley sought to make the next two mills, Granby and Olympia, bigger and better. The original charter of Richland Mills called for a firm capitalized at $150,000, and this was eventually raised to $300,000. With a structure 50 percent larger than Richland Mills, Granby Mills was initially capitalized at $150,000 and eventually raised to $800,000. The still-bigger Olympia Mills was capitalized at $1.5 million, and this, also, was later increased. While the firms were authorized to raise substantial amounts of capital, the funds never materialized in amounts sufficient to cover stock subscriptions.

28. Smith, "Economic Development," 125–42; William Elliott Memoirs, SCL.

29. Richland Cotton Mills, Application for Increase of Capital Stock to $300,000, December 19, 1899, Records of the Secretary of State, Corporate Charter Division, File #975, Charter Book P, 1–3, and Charter Book H, 96, SCDAH. See also W. B. Whaley to Olympia Cotton Mills, Deed Book AD, 321, Registrar of Deeds, Richland County, Columbia, South Carolina, hereinafter RDRC. To finance the Olympia Mills, Whaley personally purchased land and contracted for machinery, which he later signed over to Olympia in exchange for $326,000 in common stock subscriptions.

and their finances became more and more intertwined.[30] The officers and directors of Olympia served in the same capacities for the Richland and Granby mills, allowing the finances of the three companies to be tied together. Whaley and directors of the Richland and Granby Cotton mills subscribed to shares of the Olympia Mills' stock. In addition the Richland Cotton Mills Company and the Granby Cotton Mills Company both subscribed to shares in Olympia common stock.[31] To overcome continuing capital shortages, directors increased bills payable, took out loans, and opened debenture accounts. These efforts provided only temporary relief, and in July 1903, Olympia could not pay a $37,500 installment due on its debentures. Financial difficulties grew so severe that company directors decreased salaries and suspended dividend payments, and the companies' selling agents, Carey, Bayne and Smith of Baltimore, dropped the Whaley mills.[32] In October, the directors proposed a bond issue and, with stockholders' approval, the Richland, Granby, and Olympia mills issued bonds simultaneously of $450,000, $800,000, and $1,750,000. The bond issue, however, came too late to prevent reorganization.[33]

30. Smith, "Economic Development," 193.

31. Ibid., 153–59, 177–78; and Byars, *Olympia-Pacific*, 9–11. For a list of the original incorporators, see July 18, 1900, August 1, 1900, Olympia Cotton Mills, Minute Book of Stockholders' Meetings, 1899–1912, 3–22, SCL. See also W. B. Whaley to Olympia Cotton Mills, Deed Book AD, 321, RDRC. The fourth and smallest of Whaley's cotton mill ventures in Columbia received its charter on September 1, 1900, and was the only corporation of the four that did not elect Whaley president. In this way, Capital City Mills maintained financial independence from the other Whaley mills until 1904. Yet the smaller firm also suffered from chronic undercapitalization, increasing its original capitalization of $100,000 to $310,000 on June 5, 1904. See Capital City Mills, Petition for Charter, September 3, 1900, Records of the Secretary of State, Corporate Charter Division, File #2152; Return of Corporators, October 23, 1900; Articles of Incorporation, September 1, 1900; Increase of Capital Stock, July 5, 1904, all SCDAH.

32. February 13, 1902, Olympia Cotton Mills, Minute Book of Stockholders' Meetings, 1899–1912, 23–28, SCL.

33. November 14, 1903, ibid., 28–56. At this meeting stockholders of Olympia approved the issue of $1,750,000 worth of bonds. To do this, the bond issue had to be secured by a mortgage on the plan, buildings, etc., of the Olympia Cotton Mills. A similar mortgage had to be made on the properties of Richland, Granby, and Capital City mills. For a copy of the bond agreements and mortgages, see "Olympia Cotton Mills and Baltimore Trust and Guarantee Company Trust," Mortgages of Real Estate, Book AO, 169; "Granby Cotton Mills and International Trust Company of Maryland Trust," Book AO, 180; "Capital City Mills and Mercantile Trust Company," Book AO, 200; "Richland Cotton Mills and Safe Deposit and Trust Company of Baltimore Trust," Book AO, 190, all RDRC. See also Smith, "Economic Development," 193–203; and Byars, *Olympia-Pacific*, 17–19.

Reorganization brought personnel changes, a continuance of the bond issue, and centralization of the administrative functions of all four companies. Five of the seven new board members elected late in 1903 represented northern creditor interests. When Parker became president, the combined debt of Olympia, Granby, and Richland was over $1.7 million, and he moved quickly to consolidate indebtedness through the bond issue.[34] In 1910, Parker unified management of the Whaley firms in Parker Cotton Mills holding company and later sold them to Hampton Cotton Mills, an operating firm owned by Parker Cotton Mills.[35] Under new management from 1911 to 1914, the Whaley mills remained financially unstable. Operations under Hampton began with a note issue of $1 million financed by the Old Colonial Trust Company of Boston and payable in three installments: July 1, 1915, 1916, and 1917. During his presidency Parker also faced labor problems, power shortages, production curtailments, and machinery and production changes. Parker's problems in determining the most profitable cloth to produce caused further inefficiency, and each of the mills changed its production several times.[36] In addition, he invested in over 54,000 bales of raw cotton, which could not be resold and became a financial liability, tying up needed capital and costing the company over $1 million

34. Smith, "Economic Development," 203–7, 223–41; Byars, *Olympia-Pacific*, 19; Anonymous, Unbound Scrapbook of Reorganization, 1903–1905, SCL. See also William Elliott Memoirs, SCL, 24–25; William Elliott Jr. (1872–1943) Papers, 1903–1907, SCL; Whaley Mills Reorganization Committee, "Records of the Reorganization Committee," March 1–November 24, 1905, September 20, 1904, August 26, 1905, November 20, 1905, Olympia Cotton Mills, Minute Book of Stockholders' Meetings, 1899–1912, 56–90, SCL.

35. Smith, "Economic Development," 277–85. See also Hampton Cotton Mills, Petition for Charter, June 17, 1912, Records of the Secretary of State, Corporate Charter Division, File #7241; Return of Corporators, July 1, 1912; Articles of Incorporation, June 17, 1912; Olympia Cotton Mills, Certificate of Dissolution of Charter, December 27, 1913, File #1717; Capital City Mills, Certificate of Dissolution of Charter, December 27, 1913, File #2152; Granby Cotton Mills, Certificate of Dissolution of Charter, December 27, 1913, File #1038; Richland Cotton Mills, Certificate of Dissolution of Charter, December 27, 1913, File #975, all SCDAH. See also "Granby, Capital City, Richland, and Olympia Cotton Mills to Hampton Cotton Mills Company," Deed Book BG, pp. 1–8, RDRC; May 22, 1912, Olympia Cotton Mills, Minute Book of Stockholders' Meetings, 1899–1912, 104–n.p., SCL. The minutes include a resolution passed for Olympia Mills to "go into liquidation and dissolve and wind up its affairs."

36. Smith, "Economic Development," 249–77, 286–92. See also William Guion Childs (1850–1912) Papers, December 19, 1905–February 28, 1910, SCL, and William Elliott Memoirs, 25–26, SCL.

in a single year.[37] The situation grew worse when the Old Colonial Trust Company called for payment of the first $250,000 note issued to Hampton Cotton Mills in 1912. When this financial crisis led to Parker's resignation and a second reorganization, the unsecured outstanding debt of the corporation was $3.2 million.[38]

While plans were made for the total reorganization of Parker Cotton Mills, negotiations began that eventually led to the northern-based Pacific Mills' purchase of the Whaley mills.[39] In 1915 representatives of Lockwood Greene, a Boston-based engineering and architectural firm, met with M. C. Branch, the new president of Parker Cotton Mills. Branch expressed interest in hiring Lockwood Greene to manage the troubled southern holding firm, and, at a later meeting, Edwin F. Greene, president of Lockwood Greene, offered to direct Parker Cotton Mills' reorganization for a $10,000 fee. Parker Mills declined the offer but left open the option for a management contract. In February 1916, Greene and other officials of Lockwood Greene toured Parker Cotton Mills manufacturing facilities. At that time, Parker Cotton Mills functioned as a holding company for three operating firms: Monaghan Mills, Victor Manufacturing Company, and the Hampton Cotton Mills. These operating companies were individually incorporated and controlled the assets of many textile plants throughout South Carolina, with a capacity of over 500,000 spindles. The Hampton Cotton Mills, composed of the four Whaley mills in Columbia, Fairfield Mills near Winnsboro, Wylie Mills near Chester and Pine Creek and Beaver Dam Mills at Edgefield, South Carolina, operated more than 275,184 active spindles.[40] Besides being president of Lockwood Greene, Edwin F. Greene served as treasurer of Pacific Mills, and after touring the Parker plants, he informed Branch that Pacific

---

37. W. G. Childs to Lewis Parker, September 27, 1905, and Lewis Parker to W. G. Childs, September 28, 1905, William Guion Childs (1850–1912) Papers, SCL. See also Smith, "Economic Development," 295–98.

38. Ibid., 292–302.

39. Ibid., 214–20, 268, 277–81, 303–17, 330–31. In his discussion of negotiations leading to Pacific Mills' purchase of the Whaley mills, Smith utilized the transcripts of two court cases: *Heywart et al.* vs. *Parker Cotton Mills Company et al.*, in the District Court of the U.S. for the Western District of South Carolina, in Equity, 1916, and *Summersett et al.* vs. *Parker Cotton Mills et. al.*, Richland Court of Common Pleas, 1917. See also "Pacific Mills Stock Dividend," *Textile World Journal* (February 17, 1917): 8.

40. Smith, "Economic Development," 277–81, 303–4.

Mills might be interested in buying "grey goods" (unfinished textile cloth) directly from Parker Cotton Mills. This arrangement, however, rested on Lockwood Greene receiving the management contract. W. E. Beattie did not approve of contracting for Parker Mills' management and, by this time, had convinced Branch to turn down Greene's proposal. When the management contract was rejected, Greene proposed that Pacific Mills and Lockwood Greene buy the Hampton Cotton Mills and mentioned $3 million as a possible purchase price. Pacific Mills wished to buy only the four Whaley mills in Columbia; Lockwood Greene agreed to buy the four remaining mills in the Hampton group, and the Hanover National Bank of New York advanced funds for both purchases.[41]

Throughout the Whaley and Parker administrations labor policies reflected the ongoing financial instability of the Whaley mills. Accident reports and insurance claims available for the years 1900 through 1908 indicate a careless disregard of child labor laws. The mills employed children under the legal age limit of twelve years, including some as young as five or six. Reports show that both adults and children incurred serious injuries and some died from workplace accidents.[42] Though Whaley's intent had been to provide a state-of-the-art mill village, operatives' homes were constructed in a low-lying area subject to flooding, and there were outbreaks of malaria. Both Whaley and Parker complained about a shortage of laborers and the difficulty of retaining employees once they arrived in Columbia.[43] The company's refusal to provide one-time benefits or leniency in disciplinary cases prompted workers to strike in 1903 and 1907. In 1903 mill authorities refused to meet with union representatives, and Whaley fired union members and evicted them from their homes in the mill village. Reportedly, he also dispatched scouts to Georgia illegally to recruit new operatives and pay for their

41. Ibid., 304-6. See "Hampton Cotton Mills Company to Pacific Mills," Deed Book BK, 522–26, RDRC; Hampton Cotton Mills, Certificate of Dissolution of Charter, September 14, 1916, Records of the Secretary of State, Corporate Charter Division, File #7241, SCDAH.

42. Injury Reports, William Elliott Jr. (1872–1943) Papers, 1903–1907, SCL. See also Byars, *Olympia-Pacific*, 30–36.

43. Byars, *Olympia-Pacific*, 20; President's Report, November 15, 1906, Olympia Cotton Mills, Minute Book of Stockholders' Meetings, 1899–1912, 90–99, SCL. For a discussion of the "highly unstable and migratory" nature of southern textile workers, see Wright, *Old South, New South*, 142, 144; Hall et al., *Like A Family*, 105–10; and Smith, "Economic Development," 249–57.

relocation to Columbia. Whaley agreed to take back only those workers who would sign a statement saying they did not belong to and would never join a union, stating: "We are the owners of our mills and we propose to run them. . . . We do not propose . . . to have any of this unionism business." He went on to explain that mill owners throughout South Carolina had "reached an ironclad agreement not to employ union help."[44] In 1907, the weavers and loom fixers walked out of the Granby Mills when a fellow employee was discharged. The strike lasted from June until October. During that time Parker hired "special police supervision" and the general manager at Granby distributed a letter to supervisors at all the Columbia mills listing the strikers' names, thus successfully blacklisting employees participating in the strike.[45] One employee mistakenly added to the list sued Parker for punitive damages; during the trial he explained blacklisting: "There is a tacit understanding, commonly known as mill courtesy, whereby one superintendent has information from the other superintendent."[46] While Whaley and Parker struggled to secure the machinery needed to operate their factories and the financial stability needed to prevent bankruptcy, they managed labor with an iron fist, cutting costs at the expense of worker safety and disregarding child labor legislation; working with owner networks throughout South Carolina, they postponed effective labor unionization for another generation.

A lack of working capital forced engineer and entrepreneur W. B. Smith Whaley, his successor, and owners of the Whaley mills to include northern money and, eventually, northern directors in what they had originally intended to be a locally owned and operated enterprise. As Wright, Carlton, and Coclanis tell us, the Whaley mills were not unique. Whaley's experiences in Columbia were repeated throughout the South as local entrepreneurs passed control of their textile companies into the hands of northern industrialists and machinery manufacturers.[47] By 1916, when the company

44. Byars, *Olympia-Pacific*, 33–40.

45. General Manager [Granby] to W. H. Coleman, June 28, 1907, William Elliott Jr. (1872–1943) Papers, 1903–1907, SCL.

46. Byars, *Olympia-Pacific*, 38. See also J. H. Beaty, General Manager, Granby Cotton Mills, to Mr. J. S. Moore, General Manager, Capital City Mills, November 12, 1907, William Elliott Jr. (1872–1943) Papers, 1903–1907, SCL.

47. Wright, *Old South, New South*, 131; Carlton and Coclanis, "Southern Textiles in Global Context," 160–64.

purchased the Olympia, Granby, Richland, and Capital City mills, Pacific Mills was one of the largest and most stable textile firms in the world. The company easily financed the debts of the Whaley and Parker firms and placed the four modern Columbia mills into full operation. Begun through southern entrepreneurial efforts, the Whaley mills failed to compete successfully against northern manufacturers. This failure left the firm financially and technologically vulnerable and paved the way for Pacific Mills' takeover. While the relative positions of the administrators of southern and northern textile firms in relation to financial and technological networks can be seen in events leading to Pacific Mills' purchase of the Whaley mills, the power of established networks to shape the direction of a mature industry becomes even more clear when Pacific Mills' efforts to vertically integrate grey goods production and finishing in the South Carolina piedmont are considered.[48]

The Spartanburg Chamber of Commerce: Marketing Local Resources

Just over ninety miles northwest of Columbia, members of the Spartanburg, South Carolina, Chamber of Commerce worked to bring Pacific Mills to their community through a more direct route. Their efforts would result in the construction of the textile village of Lyman on a former cotton plantation near Spartanburg. On January 17, 1923, the *Greenville News* and the *State* announced the possibility of Pacific Mills purchasing land and constructing a factory near Spartanburg.[49] The following day, the *New York Times* and the *Spartanburg Herald* expanded on the announcement:

> No confirmation of the reports that the Pacific Mills Company of Lawrence, Mass., contemplates the establishing of a huge bleachery and finishing plant at Groce, twelve miles west of Spartanburg, at a cost of from $7,000,000 to $8,500,000 was available. . . . It is understood that an option of ninety days duration on the property of A. E.

48. For background information on Pacific Mills, see Donald Norton Anderson, "The Decline of the Woolen and Worsted Industry of New England, 1947–1958: A Regional Economic History" (Ph.D. diss., New York University, 1971).

49. "Great Textile Plant for Piedmont Section," *State* (Columbia, S.C.), January 17, 1923, 1; "Pacific Mills Plans Immense Bleaching Plant near Groce: New Industry May Cost over 2 Million," *Greenville News*, January 17, 1923, 1.

Groce is held by the Pacific Mills Company and that final decision on the site is to be made in the near future.[50]

It appears that the earliest negotiations to buy in Spartanburg County were carried out, as much as possible, in secrecy. The *State* based reports on "unofficial advices" concerning plans "tentatively agreed upon." The *Spartanburg Herald* reported that J. F. Kauffman, in charge of engineering work, "refused to comment," and no official statements could be obtained.[51] Alister G. Furman, a representative of Pacific Mills in the South, accused the press of false reporting and suggested that matters of industrial development be left out of public newspapers. The possibility of a purchase was not announced until almost two months of Pacific Mills' ninety-day option to buy had expired.[52] Pacific Mills' final decision was not made public until March, when the company announced plans to construct a cotton textile factory and finishing facility.[53] On March 15, 1923, a group representing Pacific Mills visited Spartanburg. The Spartanburg Chamber of Commerce arranged the day. After visiting the company's seven hundred–acre tract in Groce, they toured the nearby Pacolet Mills and Montgomery Dairy. That evening the Pacific Mills officials were guests of honor at a dinner hosted by the chamber of commerce.[54] Edwin F. Greene, speaking after the dinner, thanked Spartanburg for its warm welcome and described Pacific Mills' plans as a southern extension of the company's finishing industry.[55]

50. "Huge Bleachery Plant May Be Established at Groce," *Spartanburg Herald,* January 18, 1923, 1. See also "Pacific Mills Gets Option in South," *New York Times,* January 18, 1923, 21, col. 3; "Great Textile Plant for Piedmont Section," *State,* January 17, 1923, 1; "Plan for Large Worsted Mill Is Still Indefinite," *Greenville News,* January 19, 1923, 1.

51. "Great Textile Plant for Piedmont Section," *State,* January 17, 1923, 1; "Huge Bleachery Plant May Be Established at Groce," *Spartanburg Herald,* January 18, 1923, 1.

52. "Plan for Large Worsted Mill Is Still Indefinite," *Greenville News,* January 19, 1923, 1; "Proposed Bleachery Plant Appears to Be Assured Fact; Company Will Build," *Spartanburg Herald,* January 20, 1923, 1.

53. "Proposed Bleachery Plant Appears to Be Assured Fact; Company Will Build," *Spartanburg Herald,* January 20, 1923, 1; "Pacific Mills Will Acquire Groce Property: Plan Finishing Plant and Big Cotton Mill," *Spartanburg Herald,* March 3, 1923, 1 (actual date was February 3, 1923). See also "A. B. Groce, et. al. to Pacific Mills," Deed Book 6-T, 97–116; "Survey of Pacific Mills at Groce," Plat Book 7, 127, both Registrar of Deeds, Spartanburg County, Spartanburg, South Carolina, hereinafter RDSC.

54. "Plans for Entertainment of Textile Executives Are Complete; Here Tomorrow," *Spartanburg Herald,* March 14, 1923, 1.

55. "'Several' Millions to Be Spent at Groce," *Spartanburg Herald,* March 16, 1923, 1.

The Spartanburg Chamber of Commerce headed a community entrepreneurial effort to attract Pacific Mills to Lyman, offering incentives such as tax exemptions, inexpensive land, and an ample supply of low-wage, nonunion labor. Yet its efforts intersected with the New England firm's attempts to vertically integrate production and to streamline costs. Until 1916, Pacific Mills' dye and print works in Lawrence, Massachusetts, had to purchase grey goods from other manufacturers in order to operate at capacity. The four Whaley mills' production, combined with grey goods produced in Dover, New Hampshire, and Lawrence, brought those operations to capacity without Pacific's buying grey goods from other manufacturers.[56] The vertical integration obtained was not coincidental. The entire Hampton Mills group could have been purchased at a bargain price, but Pacific Mills bought only the mills needed to satisfy its print works capacity. Pacific Mills officials knew the capacity of their Lawrence print works and, through existing connections with Parker Cotton Mills, knew the capacity of the Whaley mills. Knowledge of Parker Cotton Mills' financial difficulties allowed the management of Pacific Mills, with the help of Lockwood Greene, to gain control of one-quarter of its ailing competitor's mills. In doing so, Pacific Mills pursued long-standing corporate policies of expansion and vertical integration while simultaneously improving manufacturing efficiency and market position.[57] This becomes more obvious when considered in conjunction with Pacific Mills' activities in Spartanburg and construction of the mill town of Lyman in 1923.

As in the 1916 Columbia purchase, the Boston-based textile engineering firm of Lockwood Greene played an important role in Pacific Mills' move into Spartanburg. During the initial stages of the purchase newspapers had trouble distinguishing the assets of Pacific Mills from those of Lockwood Greene. For example, the *Greenville News*, in a description of Pacific Mills, stated: "the company has large mills at Columbia and Winnsboro."[58] The mills in Winnsboro belonged to Lockwood Greene. Edwin F. Greene, holding positions in both corporations, carried on most negotiations in the purchase, and Lockwood Greene completed all the engineering work

56. Smith, "Economic Development," 304–5, 320.
57. Ibid.
58. "Pacific Mills Plans Immense Bleachery Plant near Groce," *Greenville News*, January 17, 1923, 1; "Boston Confirms Mill Negotiations," *Spartanburg Herald*, July 10, 1923, 1.

for the Lyman development. Lockwood Greene prepared the blueprints at offices in Spartanburg and, on April 5, 1923, the engineering firm's vice president began taking bids for the construction of Pacific Mills' plants at Lyman. Contracts for the construction of mill houses, a school, a church, a store, and needed machinery would follow.[59]

Operations began at the bleachery on May 26, 1924. By June the bleachery was putting in a kier of cloth each day and preparing to start the finishing process. The cotton mill did not begin operating until June 3, 1924. At that time the mill held 32,000 spindles and 564 looms and was capable of producing 100,000 yards of cloth weekly. The mill village also neared completion; 320 employees' homes were ready for occupants, and the community house, school building and church were being "pushed to completion." The cotton mill in Lyman could not produce enough grey goods to satisfy the new bleachery's capacity of 1.25 million yards per week, and shipments came to Lyman from other mills.[60] While this seems to contradict the policy of vertical integration established with the purchase of Whaley mills, an examination of Lockwood Greene's activities in the South Carolina piedmont proves otherwise. In August 1923, Lockwood Greene purchased Tacapau Mills, only a mile and a half below Lyman, and Pelzer Mills in Pelzer, South Carolina.[61] On September 13, 1923, the *Spartanburg Herald* reported Lockwood Greene's "acquisition of Lancaster Mills, Lancaster, South Carolina; Eureka and Springstein Mills, Chester, South Carolina; and Fort Hill Manufacturing Company, Fort Hill, South Carolina." These purchases gave Lockwood Greene over 700,000 spindles and a total of 15,000 looms in South Carolina alone. The products of these mills and those

59. "Bid for Pacific Mills Project at Groce to Be Received Here Thursday," April 4, 1923, 1; "Men to Receive Bids for Work at Groce Today," April 5, 1923, 1; "General Contract for Textile Mill at Lyman Awarded," April 11, 1923, 1; "Edwin Farnham Greene, Lockwood Greene Head, Inspects Work at Lyman," June 12, 1923, 1; "Construction at Lyman Proceeding on Schedule," August 14, 1923, 5; "Buildings for Pacific Mills Near Completion," November 15, 1923, 10; "Operations at Lyman are Attracting Many," September 13, 1923, 3; "Boston Textile Party Stops Here," December 19, 1923, 1; "Lyman Expected to Be Finished Early in Summer," January 25, 1924, 3, all in *Spartanburg Herald.*

60. "Pacific Mills Bleachery at Lyman Begins Operation," May 27, 1924, 3; "Pacific Mills Agent in City," June 2, 1924, 5; "Lyman Plant Now Ready for Work," June 4, 1924, 2; Leonard S. Little, "Lyman Development Told Of," January 16, 1924, 10, all in *Spartanburg Herald.*

61. "Tacapau Mills Sold to Lockwood, Greene and Company," *Spartanburg Herald,* August 7, 1923, 1; "Boston Interests Buy Pelzer Mills for Nine Million," *Spartanburg Herald,* August 9, 1923, 1.

in Columbia were shipped to Lyman, added to the grey cloth production there, and sent to the Lyman bleachery for finishing.[62] Not only did Pacific Mills and Lockwood Greene vertically integrate their cotton textile production in the South, they gained control of a large percentage of the textile production in piedmont South Carolina.[63]

While vertical integration, tax exemptions, and inexpensive land were used to attract Pacific Mills to the Spartanburg area, chamber of commerce members also promised plentiful supplies of low-wage, nonunion labor. Having recently experienced the Bread and Roses Strike in Lawrence and facing unionized labor in all its New England facilities, Pacific Mills gave significant consideration to labor in its decision to build a cotton factory and finishing facility in Lyman. The mill village constructed in Lyman included single-family dwellings, a school, a church, and a company-controlled store. The village and factories were constructed on a former cotton plantation and stood, though only about twelve miles from Spartanburg, in relative isolation. Lyman exemplifies the "fields to mills" company towns so often pictured in southern literature and industrial history. Dependent on Pacific Mills for jobs, schools, and supplies purchased from a corporate store, workers who came to Lyman would find it difficult to challenge employer policies. In coming to Lyman, Pacific Mills extended a network of production and a policy of integration, but the company also took advantage of a business atmosphere that favored capital over labor and a community structure conducive to corporate dominance over workers.[64]

Entrepreneurs and community leaders associated with the Spartanburg Chamber of Commerce bypassed the path to industrial development that involved competition against northern industry, opting instead to invite northern capitalists to finance and construct industry in the South. This second path to southern industrial development was followed by many community leaders throughout the southeast. Seeking to industrialize, but lacking the capital to go forward, southern businessmen and sympathetic politicians embarked

62. "Lockwood-Greene Increases Holdings in the State," September 13, 1923, 1; "Will Build Three Mills in Georgia," September 19, 1923, 1; "Lockwood, Greene Officials Coming," November 20, 1923, 1, all in *Spartanburg Herald.*

63. For a discussion of Lockwood Greene activities in South Carolina, see Carlton, *Mill and Town*, 55-60.

64. See English, "Beginnings of the Global Economy"; English, *A Common Thread.* See also Anderson, "Decline of the Woolen and Worsted."

on campaigns to attract branch factories of large national and international corporations to their towns and cities. This process attracted Pacific Mills' investment not just in Lyman but also in several Virginia locations, where the company built woolen-, worsted-, and rayon-production facilities in the 1940s.[65] Throughout the South, community entrepreneurs and political leaders sought to bring industry and jobs to their communities by whatever means necessary, but in almost every case attracting extraregional capital and technology to their communities was essential. In these cases, regional industrial development occurred through the extension of financial, technological, and business networks based in New England, like those of Pacific Mills and Lockwood Greene executives, into a less developed region.[66] Firmly established and financially secure, Pacific Mills could exploit regional cost advantages without the worries of capital formation, stock subscriptions, or machinery costs faced by many southern industrialists, such as Whaley and his associates. The company could also exploit regional probusiness political networks that worked to prevent unions and to limit legislation that might benefit or protect workers.

From Graham to Carr in Carrboro:
Successful Southern Competition

In 1895, George M. Graham, Paul Cameron Graham, and George P. Collins pooled their resources and opened the Durham Hosiery Company in Carrboro, North Carolina. George M. Graham, as secretary and treasurer, accepted responsibility for day-to-day

65. "Novel Arrangement Made to Provide Power for Worsted Mill Plant Located at Halifax," *Halifax Gazette* (South Boston, VA), November 7, 1946, 10; "Brookneal Is Selected after Careful Consideration," *Union Star*, May 9, 1947, 1; Josef Berger, ed., *Memoirs of a Corporation: The Story of Mary and Mack and Pacific Mills* (Boston: Barta Press, 1950), chap. 9, 15–17; "Resolution," *Charlotte Gazette*, February 16, 1948, 2; "Town Entertains Mill Officials at Luncheon," *Charlotte Gazette*, March 25, 1948, 1. For additional information on Pacific Mills, see Anderson, "Decline of the Woolen and Worsted."

66. For a discussion of efforts to attract industry to communities and the personal networks allowing it to take place, see Carlton, *Mill and Town in South Carolina*, 40–81. See also English, "Beginnings of the Global Economy," 175–98. English discusses the successful efforts of Alabama entrepreneurs and politicians to convince the Dwight Manufacturing Company, a Massachusetts textile firm, to construct a factory in Gadsden, Alabama.

management of the new enterprise. With little managerial or manufacturing experience, he struggled to keep the company solvent. Constantly in need of capital, dependent on northern machinery manufacturers and selling agents, and unable to hire skilled operatives in the rural South, he faced no easy task. After three frustrating years, during which he managed to get little hosiery to market, Graham admitted defeat and sold controlling interest in the company to another local textile entrepreneur.[67] In the same year that Graham began his hosiery venture, Julian S. Carr Sr. opened the Golden Belt Hosiery Company, also near Carrboro. Though Carr faced the same difficulties as Graham, his prior financial success with tobacco and banking enterprises kept Golden Belt alive, and in 1898, he purchased the Durham Hosiery Company. Carr merged the two companies at an opportune time, incorporating them as Durham Hosiery Mills.[68]

Like Whaley in Columbia, George W. Graham, as head of Durham Hosiery Company, encountered problems with purchasing, receiving, installing, and maintaining production machinery. This task proved especially difficult for southern manufacturers because the equipment they purchased had to be shipped from New England and serviced by companies based in New England. Poorly constructed or flawed equipment could not be easily returned for repair; bringing technicians from the manufacturer to the factory for the repairs was expensive and difficult to arrange. Many machinery manufacturers designated only one sales representative for the entire South, creating a constant shortage of knowledgeable machinists.[69]

---

67. George M. Graham Letterbooks, 1895–1898, Durham Hosiery Mills Papers, 1887–1962, Rare Book, Manuscript, and Special Collections Library, Duke University, Durham, North Carolina, hereinafter DHM. See also Jean Bradley Anderson, *Durham County: A History of Durham County, North Carolina,* (Durham: Duke University Press, 1990), 213; Mena Webb, *Jule Carr: General without an Army* (Chapel Hill: University of North Carolina Press, 1987), 176–77; William Kenneth Boyd, *The Story of Durham: City of the New South* (Durham: Duke University Press, 1925), 124.

68. Anderson, *Durham County,* 213; Webb, *Jule Carr,* 175–77; Boyd, *The Story of Durham,* 127; Manuscript Drafts of Stockholders' Meeting Minutes, January 22, February 9, 1898, Letters and Papers, 1898–1899, Folder 5, Box 20-G, DHM Records. For a discussion of the early industrial development of Durham revolving around the tobacco and cotton textile industries, see Boyd, *The Story of Durham,* 51–123; Anderson, *Durham County,* 138–152, 172–190; and Webb, *Jule Carr,* 70–85.

69. See series of letters between Graham and Providence Knitting Machinery Company, Providence, Rhode Island, including September 23, 1895, September 26, 1895, October 4, 1895, George M. Graham Letterbooks, 1895–1898, DHM.

The hosiery industry faced particularly difficult technical problems. By the 1890s, knitting machinery had advanced to the point that unskilled operatives could easily tend correctly calibrated machines. But the machines required a skilled technician to keep them correctly calibrated and balanced; otherwise, broken needles slowed production to a crawl. In the four years that George W. Graham owned and managed the Durham Hosiery Company, he became a jack-of-all-trades. Graham ordered, purchased, and helped to install all machinery. When the knitting machines failed to work as anticipated, he carried on a relentless correspondence with the northern-based manufacturers, requesting aid and replacement equipment. In one instance, Graham explained to the Providence Knitting Machinery Company from which he had purchased fifteen knitters: "The machines we have are running terribly. . . . The plain truth though is this—you are [sic] in such a hurry to fill our order that the machines were not built right. If you wish to keep up the reputation you have made in this state for honest dealing you will take these machines back."[70] Graham also worked directly with unskilled laborers, teaching them how to operate the looms and loopers, and, when quality-control problems continued, he attempted to recruit skilled foremen from New England.[71]

Shipping textile equipment between machine shops and factories within New England could be cumbersome, but not nearly as complex as moving equipment from Massachusetts, Connecticut, or Rhode Island machine shops to North Carolina or back from North Carolina for repair. Going to and from the South required that machinery and goods change hands, boats, or trains several times before reaching their final destinations. Security and careful handling could not be assured, and returning an item or shipping it back for repairs became a costly endeavor. The Durham Hosiery Company experienced continuous problems with machinery shipments, including lost orders, wrong orders, and items damaged in transit. Writing the Union Special Sewing Machine Company of New York, Graham acknowledged receipt of a shipment but asked, "Why did you not send the machine we ordered? You sent no speed

70. Graham to Providence Knitting Machinery Company, September 23, 1895, Graham Letterbooks.

71. Graham to H. W. Bigelow, Esq., Providence, RI, June 14, 1895; Graham to C.O. Bean, July 17, 1985, both Graham Letterbooks.

cone for driving machine; no foot sewer shifter nor any guard attachment for concealed stitch for welting tops of ladies hose as we expressly wrote you we wanted."[72] In another instance, he explained to James Taylor of Philadelphia: "The looper & belt hooks just came. Upon opening the box we find that it was so insecurely packed that several of the points were broken & others mashed against the side of the box."[73] When machinery orders arrived undamaged, correct installation and calibration often proved difficult. In April 1896, Graham wrote to the Mayo Knit Machinery Company in Franklin, New Hampshire, explaining that cog wheels on recently installed knitting equipment "have broken & some 3 or 4 times. It costs us a good deal to get them fixed here & besides the loss of time & work."[74] When technical difficulties could not be solved in-house, Graham was forced to request that northern machinery manufacturers send engineers or machine fixers to help solve production problems. In July 1896, Graham asked the Putnam Machine Company to send an engineer to work with recently installed loopers, explaining "We will not need him over 2 days."[75] On several occasions, Graham abandoned attempts to fix malfunctioning equipment and shipped it back to the manufacturer.[76]

Graham attempted to avoid these problems by carefully articulating his technical needs to manufacturers and enclosing samples or diagrams. For example, he requested that the Keystone Knit Manufacturing Company of South Bethlehem, Pennsylvania:

> Be sure that the needles are made exactly like the needles . . . we enclose. Our machines allow of no variation & the 1/64 of an inch variation in the length of the shank would cause that machine to break a great many more needles. . . You will find enclosed another sample with a curve just above the shank. The curve helps the band to hold the needles in place. Can you make yr [sic] spring bottom needles with that curve?[77]

72. Graham to Union Special S.M. Co., N.Y., August 22, 1895; Graham to Prov. Knitt Mach. Co., Providence, RI, August 26, 1895, all Graham Letterbooks.
73. Graham to James Taylor, September 23, 1895, Graham Letterbooks.
74. Graham to Mayo Knit Mach. Co., April 15, 1896, Graham Letterbooks.
75. Graham to Thomas K. Carey and Bros. Co., July 9, 1896, Graham Letterbooks.
76. See for example, Graham to Mr. R. Berry, Providence, November 12, 1895, Graham Letterbooks.
77. Graham to Keystone Knit Mch. Mfg. Co., So. Bethlehem, March 28, 1896, Graham Letterbooks.

When purchasing bigger and more expensive items, Graham attempted to arrange to take them on a trial basis, delaying payment until performance quality could be evaluated.[78] Despite these efforts, between 1895 and 1898, Graham constantly dealt with problems caused by poorly constructed or damaged machinery, problems enhanced, if not caused, by poor communication and transportation services.

Transportation and communication were no more reliable going in the opposite direction. Southern firms were often forced to send their products north to be finished, dyed, or marketed (or all three). Graham purchased cotton yarn from all over the South, designed and ordered packaging, and carried on unending negotiations with northern dye companies to achieve the perfect "fast black" and a rapid turnaround for his hosiery.[79] Marketing quickly became one of Graham's most difficult problems. His dependence on anonymous selling agents in New York City found expression first in subservient acquiescence to expressed opinions as to the quality of his product and later, as he grew more knowledgeable, in sarcastic denial of their abilities to sell anything. When he first attempted to market his product, Graham established a relationship with a New York agent, Joseph Cotler. Early on, Graham seemed almost apologetic in his dealings with Cotler, repeatedly thanking him for his "criticisms" and promising "I will do my best to remedy the faults."[80] After several frustrating months, Graham's tone began to change: "If the goods were not right you should have written us before. . . . I sincerely hope you will dispose of all the goods you have on hand."[81]

Until the Carr family took over, the hosiery industry throughout the North Carolina piedmont faced many problems associated with its dependence on northern machinery manufacturers and business, financial, and marketing networks. Julian S. Carr Sr. grew up

78. See for example, Graham to Providence Knitting Machine Company, Providence, September 23, 1895, Graham Letterbooks.

79. See for example, Graham to Savannah Cotton Mills, Savannah, Georgia, February 3, 1896; Graham to Mr. Joseph Kahn, Atlanta, September 24, 1895; Graham to Franklinville Dye Works Co., Philadelphia, August 15, 1895; Graham to Franklinville Dye Works Co., Philadelphia, October 15, 1895; Graham to Continental Dye Works Co., Philadelphia, October 17, 1895, all Graham Letterbooks.

80. Graham to Mr. Jas. Colter, New York, September 27, 1895, Graham Letterbooks.

81. Graham to Colter, March 11, 1896, Graham Letterbooks.

the son of a prosperous merchant, attended the University of North Carolina, fought in the Civil War, and invested heavily in the industrial development of the Carolina piedmont after the war. He had had success with his Bull Durham chewing tobacco, and few doubted Colonel Carr's ability to move successfully into hosiery production. Carr's success in business was built on his investment in transportation, communication, financial, technological, and political networks. Carr, along with Eugene Morehead, W. T. Blackwell, E. J. Parrish, Brodie Duke, Benjamin Duke and other successful piedmont Carolina businessmen, invested in and served on the boards of banks, railroads, newspapers, telegraph, water and electricity companies, and hospitals. Comparable in kind, if not in scale, to the overlapping investments and boards controlled by the Boston Associates in New England in the early nineteenth century, the Piedmont Carolina Associates built regional networks that sustained industrial development in and around Carrboro, Durham, and Raleigh.[82]

Beyond his business networks, Carr was active in local, state, and regional politics. He ran for public office, directed the editorial voice of newspapers he owned, and supported the Democratic Party agenda. While noting that Carr was a Democratic candidate for senator in 1900, biographer Mena Webb explains that "ultra-conservatives" in the Democratic Party "opposed Carr because he favored election reforms, free silver, and state-supported education," and during the campaign they "accused Carr of attacking the means by which white supremacy was attained in 1898" and of "supporting Populists and Republicans against 'good Democrats.'"[83] Carr defended himself against these attacks by pointing to his financial contributions to the North Carolina economy, public and private education, and the Democratic Party.[84] But Carr's contributions to the party went beyond financial support; he actively supported the party's white supremacist agenda. In the 1898 campaign, he traveled extensively

82. For background and biographical information on Carr, see Webb, *Jule Carr*, 70–103, 176–190; *Dictionary of North Carolina Biography*, s.v. "Carr, Julian Shakespeare"; and Anderson, *Durham County*, 213. For a discussion of the development of the financial, transportation, and communication networks that supported the industries of Durham County, see Anderson, *Durham County*, 183–90; and Boyd, *The Story of Durham*, 115–23.

83. Webb, *Jule Carr*, 168–75.

84. Ibid., 168–75. See also Anderson, *Durham County*, 231, 261.

within North Carolina speaking in support of a state constitutional amendment that ultimately disfranchised black citizens and denouncing "black rule" in the state.[85] Carr not only participated in the campaign to disfranchise blacks, he went about the state and throughout the country defending white supremacy, southern honor, the Lost Cause, the security of white female virtue, and lynching. Carr helped establish more than a network of institutions and technology that supported industries; he actively sought and supported racial segregation in the South, approved of the violence of that system, and used his ties with communication, transportation, financial, and religious networks to help sustain it.[86] While Carr's support for the Democratic Party's white supremacist agenda cannot be disputed, in Eric Anderson's *Durham County* we learn that Carr gave his money and his time to the construction and success of black schools, churches, colleges, and businesses. While defining Carr as a radical white supremacist, Anderson also notes that both W. E. B. Du Bois and Booker T. Washington, upon visiting Durham, praised Carr for his support of the city's black community.[87] In other words, Carr supported black institutions as long as they were segregated institutions that did not challenge the racial or labor status quo. Julian S. Carr Sr. was a highly successful and gifted entrepreneur, a generous philanthropist, a man of power and prestige, seemingly respected by both elite and working-class whites throughout North Carolina. Yet, in Carr, we can document the link between the networks of politics, finance, transportation, communication, and technology upon which southern industry was built and the networks by which economically and politically powerful white men controlled labor and race relations throughout the South. This link in the person and personal network of Julian Carr between white entrepreneurial actions and white supremacist

85. See Helen G. Edmonds, *The Negro and Fusion Politics in North Carolina, 1894–1901* (Chapel Hill: University of North Carolina Press, 1951); Eric Anderson, *Race and Politics in North Carolina 1872–1901: The Black Second* (Baton Rouge: Louisiana State University Press, 1981).

86. See correspondence and collection of speeches delivered by Julian Carr Sr., between ca. 1890s–1930s, General Papers, Carr Papers, #141 in the Southern Historical Collection, Manuscript Department, Wilson Library, The University of North Carolina, Chapel Hill (SHC). See also Anderson, *Durham County*, 260–61.

87. Anderson, *Durham County*, 156-63, 222, 231, 257-61. Durham was also a center of black business success. For a discussion of late-nineteenth-century black business developments in the town, see pp. 220–24.

politics must inform our exploration of managerial and business development of the Durham hosiery industry in the 1890s and the early 1900s.

Hoping to bypass the fledgling southern industry's dependence on northern selling agents, in 1900 Carr established his own commission firm in New York, the Carolina Hosiery Company. The commission agency served as selling agent for several hosiery firms in the Southeast, but particularly the Durham Hosiery Company. Finding strong markets, Carr increased production, purchased new machinery, updated old facilities, and built new factories.[88] With A. T. Bloomer, general manager of the Carolina Hosiery Company, serving as selling agent and product advocate, things improved for the hosiery manufacturers of the North Carolina piedmont. Working out of New York, Bloomer sold more hosiery for the North Carolina manufacturing firm than had previous agents. He also helped to coordinate hosiery production among the region's various manufacturers and advised them on what products to manufacture, which machinery to use, and how to improve the quality of mill products. The success brought by the selling agency he established placed Carr Sr. in a position to initiate consolidation of hosiery manufacturing in North Carolina.[89]

As president of Durham Hosiery Mills, Carr practiced family ownership and management. He brought his relatives into the hosiery business with him, including his sons, Julian S., Jr., Claiborn M., and Marvin, and his nephew, William F. Carr. Company records also show an A. H. Carr and a W. A. Carr serving as managers and a C. A. Carr serving as a department foreman. These Carr family members owned stock in the company, and at least one Carr served as superintendent or overseer at each factory operated by the corporation.[90]

88. Ibid., 212–13; Webb, *Jule Carr*, 177–78.

89. See A.T. Bloomer correspondence with the Durham Hosiery Mills in Letters and Papers, 1898–1899, Durham Hosiery Mills Papers, Box 20-G, Folders 2–5, DHM, and Manuscript Drafts of Stockholders' Meeting Minutes, January 22, 1898 and February 9, 1898 in Letters and Papers, 1898–1899, Durham Hosiery Mills Papers, Box 20-G, Folder 5, DHM. See also Webb, *Jule Carr*, 177; and Anderson, *Durham County*, 213.

90. See Manuscript Drafts of Stockholders' Meeting Minutes in Letters and Papers, 1898–1899, Durham Hosiery Mills Papers, Box 20-G, Folders 2–5, DHM; and Minutes, Senate and House Employee Representative Meetings, 1919–1921, in Durham Hosiery Mills Papers, DHM. See also Webb, *Jule Carr*, 176–90; Anderson, *Durham County*, 241–42.

The conditions under which the first operatives in the Durham hosiery industry labored are obscure. We do know that the majority of hosiery mill workers were white and female, and sources indicate that in 1896 and 1897, operatives at the Durham Hosiery Company worked "from sixty-eight to seventy hours a week" and "made about four cents an hour."[91] The fact that Julian S. Carr Sr. began construction of new mill villages almost immediately after merging the Golden Belt and Durham Hosiery enterprises in 1898 suggested that earlier housing and living arrangements were considered inadequate. Carr Sr.'s management of laborers was paternalistic. The Durham Hosiery Mills owned the workers' houses, churches, stores, and schools. In addition, the labor force was organized around nuclear families, and in some families, several generations worked for the Durham Hosiery Mills.[92]

Late in 1900, Julian S. Carr Sr. retired, and his son and namesake, Julian S. Carr Jr., became president of Durham Hosiery Mills. The younger Carr continued to utilize family ties and to build upon his father's success. But rapid expansion—not only in sales, but also in manufacturing facilities—made it impossible to maintain a system of hands-on family management.[93] In 1903 and again in 1919, Carr Jr. opened new factories that employed only black operatives. This went against accepted racial policies of the time, and for "this he met with prejudice and criticism" and even threats of violence from white workers.[94] In 1909, the company purchased two mills in

91. *Durham Globe* (newspaper), December 12, 1895; see also Time Books, 1895–1897, in Durham Hosiery Mills Papers, DHM.

92. See Senate Minutes, October 5, 1920, Senate and House Employee Representative Meetings, 1919–1921, in Durham Hosiery Mills Papers, DHM. See also Webb, *Jule Carr*, 176–90, 243–46.

93. Manuscript Drafts of Stockholders' Meeting Minutes, 1919–1921, in Letters and Papers, 1898–1899, Box 20-G, Folder 5, Durham Hosiery Mills Papers, DHM; Anderson, *Durham County*, 241; Webb, *Jule Carr*, 176–90. While Boyd in *The Story of Durham* states on page 124 that Carr Sr., immediately after the 1898 merger, engaged a "Mr. P. Sheridan, who had been engaged in the hosiery business in New England," no confirmation of this information has yet been identified. In addition, while Sheridan may have served as a "new manager," he clearly did not direct the Carr hosiery enterprise after 1900 when Julian Carr Jr. and his other sons took control of the family-owned company.

94. Anderson, *Durham County*, 241–42. See also Boyd, *The Story of Durham*, 125–26. Despite his determination to employ black workers in his hosiery mills, in a 1919 article, Carr Jr. voiced paternalistic attitudes toward black laborers, and to some degree white laborers, similar to those of his father. See also Julian S. Carr Jr., "Building a Business on the Family Plan," *System* (July 1919): 47–50.

Chapel Hill, Alberta Mills from Thomas Lloyd and Blanche Mill from the Pritchard brothers and W. E. Lindsey. One year later, additions to the original Golden Belt Mill increased employment there to over seven hundred operatives. In 1912, the Carrs added a spinning mill on Elm Street to manufacture yarn for the hosiery mills. According to Jean Bradley Anderson, "Even before this addition the Durham Hosiery Mills were reported to be the largest in the world, employing 950 workers, mostly women and girls. The new mill was expected to add 150 to 200 more."[95] In 1918, the company "enlarged" North State Knitting Mills "from 85 to 450 machines capable of making up to 2,500 pairs of white socks a day."[96] By 1921, Durham Hosiery Mills included the two original factories, plus fourteen additional manufacturing sites in and around the Durham area. Besides the lower-quality children's footwear, the mills produced high-quality "Durable Durham" women's cotton and silk hose and men's cotton socks.[97]

Between 1900 and 1920, the Carr brothers "converted a small, mortgaged knitting mill into the nucleus of a $6 million enterprise."[98] During this era of growth Durham Hosiery Mills also experienced changes in management style and corporate culture. In some ways these changes seemed unavoidable, forced by the very pace of expansion and increased production. When he became president, J. S. Carr Jr. continued his father's practice of placing a member of the Carr family as superintendent at each mill. He soon found himself spending most of his time in New York, overseeing sales through the Carolina Hosiery Company. His brother Claiborn continued to reside in Durham for some time. By 1920, he had begun to travel more frequently between New York City and North Carolina. Records indicate that mill superintendents reported activities to Carr Jr. in New York, first by telegram and later by telephone. As expansion continued, however, this process proved inadequate. With sixteen separate facilities scattered throughout the Carolina piedmont, an office and selling agency to maintain in New York

95. Anderson, *Durham County,* 241–42; Webb, *Jule Carr,* 176–90.

96. Anderson, *Durham County,* 242.

97. Ibid., 241–42, 524 n. 9. For similar discussions of the expansion of Carr hosiery enterprises see Boyd, *The Story of Durham,* 124–27; and Webb, *Jule Carr,* 176–90.

98. Anderson, *Durham County,* 242.

City, and (until the depression) an ever-increasing demand for their product, Durham Hosiery Mills could no longer rely solely on family management.[99]

In addition to their expanding industrial enterprise, the Carr brothers found Durham Hosiery Mills susceptible to labor unrest. Between 1900 and 1915 "a new wave of union organizing hit" the Durham area, and Carr Jr. responded to labor challenges with a company-controlled social welfare program, expanding the responsibilities of his personnel department. He hired a nurse for mill village residents; instituted a profit-sharing plan; constructed a park and playground; provided a company newspaper, band, and baseball team; created health and life insurance funds for older employees; and set up a company credit union for employee loans. A night school, added in 1915, provided classes for any employee who wanted to attend.[100] But hosiery workers refused to be bought, and in 1916, a strike for higher wages at Golden Belt Hosiery and Chatham Knitting Mills forced Carr Jr. to increase wages by 7.5 percent, reduce the workweek by two hours, and lengthen the lunch break by twenty minutes in all his mills.[101] It was also in 1916 that Carr Jr. added cost and production departments to the existing personnel department. In so doing, he created a middle, or managerial, level between the company owners and the factory workers. The company gradually became an establishment with absentee owners, operated on a daily basis by trained managers.[102] While clearly tied to labor unrest, in part, the administrative changes Carr Jr. and his brothers made reflected the trends of their time; in the first two decades of the twentieth century, owners and managers in many U.S. industries sought to improve administrative efficiency and to prevent labor unions.[103] The stated purpose of Durham Hosiery

99. See Correspondence, 1900–1919, Durham Hosiery Mills Papers; and Minutes, Senate and House Employee Representative Meetings, 1919–1921, both DHM. See also Webb, *Jule Carr*, 176–90. For a discussion of Carr Jr.'s approach to management and direction of the Durham Hosiery Mills, see Carr Jr., "Building a Business," 47–50.

100. Anderson, *Durham County*, 243–46.

101. Ibid., 243–44.

102. See Correspondence, 1900–1919, Durham Hosiery Mills Papers; and Minutes, Senate and House Employee Representative Meetings, 1919–1921, both in DHM.

103. See Webb, *Jule Carr*, 176–90; Anderson, *Durham County*, 242–46. See also Irving Bernstein, *The Lean Years: A History of the American Worker, 1920–1933* (Boston: Houghton Mifflin Co., 1960).

Mills' personnel department was to serve as mediator between management and workers. The department helped to carry out social programs in the mill villages and benefit programs for factory employees. The cost department determined expenditures for each production process and generated cost reports, identifying areas for possible improvement. W. M. Upchurch reported that in the first six months of operation, the cost department saved the company over forty thousand dollars. The production department responded to the findings of the cost department, directing foremen in the installation of new, more efficient procedures and equipment.[104]

Compromise and wartime prosperity eased labor pressures for a while, but in 1919 efforts to unionize Durham's textile workers were renewed. The International Textile Workers Union, calling for reduced hours and a minimum age of sixteen years for factory workers, succeeded in recruiting over eight hundred members in less than eight months. By 1920, Durham workers boasted of a new central labor union formed with representatives from various locals.[105] Not surprisingly, Carr Jr. chose this moment to introduce an employee-representation program, or company union, which allowed employees to express opinions and present ideas in meetings controlled by the corporation. Dubbed "Industrial Democracy," this form of company union was embraced by many American manufacturing firms after World War I, including other textile firms in the surrounding area. The company union was patterned after the U.S. government with a house of representatives, a senate, and a cabinet. As head of the personnel department, W. W. Shaw repeatedly reminded the representatives and senators that the personnel department remained "neutral," supposedly taking no sides on any issue.[106] House members, workers elected by fellow operatives, represented the interests of their particular departments or production areas and carried workers' concerns to factory management. Foremen and higher-ranking employees from each manufacturing department became members of the senate, and the cabinet level, or executive

104. Minutes, Senate and House Employee Representative Meetings, 1919–1921, Durham Hosiery Mills Papers, DHM.

105. Anderson, *Durham County*, 313.

106. House Minutes, December 1, 1920, Senate and House Employee Representative Meetings, 1919–1921, Durham Hosiery Mills Papers, DHM. For a general overview of the "Industrial Democracy" labor program established in the Durham Hosiery Mills see Anderson, *Durham County*, 312–16; Boyd, *The Story of Durham*, 131–34; Webb, *Jule Carr*, 185–87.

branch, included all managers and owners, among them many members of the Carr family. Managers could not be members of the house or senate, and the heads of the cost and personnel departments attended the meetings of all three branches. Carr Jr. adapted this labor program from John Leitch's *Man-to-Man: The Story of Industrial Democracy* and referred to Durham Hosiery Mills' company union as "industrial democracy."[107] In establishing a company union for Durham Hosiery Mills, Julian Carr Jr. sought to achieve four goals: to enhance the company's public image; to cut costs by making the production process more efficient; to improve communication between management and employees; and to prevent labor unrest and unionization. Employees of Durham Hosiery Mills, while initially cooperative, ultimately rejected the company union, and by the end of the decade, they had once again attempted to establish a local union linked to national and international labor organizations.

Once in control of the hosiery industry in the North Carolina piedmont, the entrepreneurial Carr family successfully established regional financial, business, communication, transportation, marketing, technology, and labor-control networks over which they had some degree of dominance. Because they concentrated on hosiery production, a specialized field of textile manufacturing, and established more adequate supplies of local capital, the Carr family gained a greater degree of independence from northern networks than did the Whaley mills in Columbia or the chamber of commerce in Spartanburg. Thus, Durham Hosiery Mills was able to compete successfully against northern hosiery manufacturers throughout much of the twentieth century. Not until the 1940s and 1950s would Pacific Mills arrive in Carrboro and then only to take over mills recently abandoned by Durham Hosiery Mills for newer facilities. While Durham Hosiery Mills' success rested partly on the entrepreneurial efforts of Julian Carr Sr. and his sons, it also rested on his links to a network of southern politicians and businessmen who worked together to control labor, both white and black. In the early twentieth century, textile operatives in Carrboro and throughout the Southeast challenged the labor-control networks of southern

107. John Leitch, *Man-to-Man: The Story of Industrial Democracy* (New York: B. C. Forbes Co., 1919). For additional information on the company union movement in the United States, see Bernstein, *The Lean Years*, 159-69; and Wright, *Old South, New South*, 147–55.

industrialists when they sought to form unions and to unite with national and international labor organizations. Strikes in 1916, unionization in 1919, and regionwide labor unrest in 1929 and 1934 all represent southern laborers' efforts to access existing and mature organized unions and labor networks initially established by textile workers in the Northeast.

Conclusion

By the 1880s and 1890s, the northern textile industry had undergone dramatic changes in business and financial organization, management structure, and labor relations. As the South entered its industrial revolution, it did so at a time when the business, financial, technological, and labor networks of the New England textile industry were taking on extraregional characteristics. Marketing and financial institutions of the industry were centered in New York. The exchange of technology and technological knowledge took place on local, regional, extraregional, and (to some degree) international levels. To enter the textile industry, southern entrepreneurs, businessmen, and community leaders had no choice but to acknowledge those existing networks. To bring textile factories to their communities, southern entrepreneurs not only had to compete against northern textile companies, but also had to seek entry into northern business, financial, and technological networks that were controlled by the northern branch of the industry. Southern entrepreneurs followed a variety of paths; no one method of industrial development shaped southern regional industrialization. Rather, entrepreneurs in each southern community sought out the most likely avenue to successful industrial development for their particular community. In Columbia, Whaley used his personal connection with New England textile and textile machinery manufacturers to finance, construct, and bring the four Whaley factories into production. When local capital could no longer sustain the operations, northern textile interests were so heavily involved that Pacific Mills' takeover of the firms seemed a mere formality with little change in top personnel. In Spartanburg, local entrepreneurs invited and enticed Pacific Mills executives to purchase land and construct new manufacturing and finishing facilities. In Carrboro, the Carr family developed relationships with other southern entrepreneurs to control banks, railroads, and marketing networks within the Carolina piedmont that allowed them to compete successfully in hosiery production, a less developed niche in the textile industry. The New South did not develop by following any single path to industrialization. To the contrary, in community after community, southern entrepreneurs, through failure and success, worked out a variety of accommodations with the long-established northern business, financial, and technological networks against which they were forced to compete.

In all three case studies presented here we see the extension of an existing and mature capitalist system—a system characterized by mature financial, business, and technological networks—into a previously underdeveloped region of the country. This envelopment of previously underdeveloped regions by an expanding capitalism occurred through the interaction of individuals in banks, factories, engineering schools, machinery factories and on boards of directors in both regions. New South industrial development depended on entrepreneurs in local communities gaining access to the resources in capital formation, business organization, marketing, or technological knowledge and resources controlled by those already in association with national and international economic networks. The textile industry stood at the forefront of southern industrialization in the late nineteenth and early twentieth centuries. The process by which southerners established textile factories in their communities differed from town to town but tended to involve some degree of compromise with the northern branch of the industry. This variety in local industrial development from community to community in the New South illustrates an underdeveloped region's need to accommodate itself to broader and more established networks within a mature capitalist or industrial economy. Rather than following any clearly defined process of industrial development, business and political leaders weighed the relative assets and liabilities of their communities and then proceeded to industrialize through what appeared to be the most rapid means at their disposal. In every community, however, access to capital, technical knowledge, or markets, and the question of how to obtain that access were essential. Perhaps underdeveloped regions do not industrialize as regions at all; rather, underdeveloped regions industrialize community by community and on a local entrepreneurial level. Community leaders may seek support in state legislatures, may seek association with regional organizations, may copy their neighbors' successful tactics, but, ultimately, local entrepreneurs and community leaders must find a way to access the existing financial, technological, and marketing networks that will bring industry to their towns and jobs to their citizens.

One network, however, stands apart in the three case studies presented here, and that is the labor network. Labor is always a dual network because both owners and laborers themselves seek to control it, or at least to control some part of it. Using Whaley in Columbia, chamber of commerce members in Spartanburg, and both Carr Sr. and Carr Jr. in Carrboro as case studies in southern entrepreneurial labor relations, a clear pattern emerges of entrepreneurial domination of southern labor. While they were not all successful in establishing regional technology, business, and financial networks, southern textile mill entrepreneurs and owners created an antilabor network that kept wages low and both working and living conditions unregulated. Their antilabor network used blacklisting, company unions, and harsh strikebreaking methods. The network exerted control over the press, segregated white and black workers, and influenced membership in the legislative bodies that might vote to regulate working conditions. Underpinning the network was mill owners' ability to isolate workers in rural textile villages, preventing them from gaining access to labor organizers from established unions. Despite this, textile mill workers in both Columbia and Carrboro sought and sometimes gained access to a broader network of labor organization and agitation. They participated in successful strikes, forced concessions from mill owners and administrators, joined unions and formed locals, defended fellow workers from abuse, refused to participate in company unions and, in Columbia in the 1930s, negotiated a labor contract with Pacific Mills executives. Just as entrepreneurs in underdeveloped regions must gain access to existing networks within the mature capitalist economy, workers in underdeveloped regions must have access to existing and mature labor networks, organizers, national and international unions, lobbyists, supportive politicians, and political parties. Without that access, industrial workers find themselves at the mercy of those who would sacrifice them for economic development, putting their safety as well as their economic and political interests at risk.

Technocracy on the March?

The Tennessee Valley Authority and the Uses of Technology

STEPHEN WALLACE TAYLOR

AMERICAN ENGINEERS IN THE EARLY TWENTIETH CENTURY WERE in many ways the truest progressives, for they believed above all in progress—they possessed a faith that human reason, guided by experience and careful application of scientific method, could solve or at least alleviate most of the problems faced by society. That faith is at the core of the idealism that characterized the early years of the Tennessee Valley Authority (TVA), and it is also at the heart of the ossification and bureaucratization that plagued the agency in the 1960s. It may also be central to an understanding of the agency's exploration of new technology in the late 1970s and early 1980s.

Engineering was originally a purely hands-on discipline, and engineers were not generally trained in the classroom; rather, they were called mechanics, and they trained on the job, as apprentices. Engineering became professionalized at about the same time as other professions, and for many of the same reasons—professional engineers trained in classrooms had greater credibility and greater autonomy because of their exposure to the latest knowledge, while mechanics trained on the job ostensibly knew only the things required for their particular work environments. Thus, American engineers in the early twentieth century possessed an independence of thought and an elitist bent that, taken together, could lead away from democracy and toward technocracy. As one critic explained the logic of technocracy:

Since professional work cannot be understood fully by outsiders, the person in charge of such work should be a member of the profession. In this manner, doctors have insisted that the heads of hospitals and medical schools should be members of the medical profession. . . . In the case of engineering, this principle can be extended much further. Engineering is intimately related to fundamental choices of policy made by the organizations employing engineers. This can lead to the assertion that engineers ought to be placed in command of the large organizations, public and private, which direct engineering. This is tantamount to saying that society should be ruled by engineers.[1]

An awareness of this tendency on the part of managers or elected officials might check such elitism, but there could be no guarantees that a less informed public would understand, or approve, decisions made by those professionally trained to make them. Indeed, many engineers disdained "politics" as a means through which sound technical decisions might be undermined or reversed for personal gain. In situations where infrastructure cost was large, as with electrical power, government financing might prove very important, but the involvement of elected officials would seem at best inconvenient.

Thus we can see the collision between engineers and the political process by examining the electric power industry. Demand for electric power increased tremendously in the early 1900s, with the number of U.S. homes wired for electricity growing from 8 percent in 1902 to 34.7 percent by 1920 and electric machinery replacing that for steam in industrial settings. Production grew rapidly as well, with a nearly fourfold increase over the same period. Power company officials used the increasing demand to justify calls for increased production capacity and increased interconnection among the nation's major power companies. One such proposal for interconnection was the Super Power plan, first proposed by William S. Murray in 1919 and later endorsed by Secretary of Commerce Herbert Hoover as a solution to fluctuating demand and varying efficiency of the electric power supply along the corridor from Boston to Washington, D. C.[2]

1. Edwin T. Layton Jr., *The Revolt of the Engineers: Social Responsibility and the American Engineering Profession* (Baltimore: Johns Hopkins University Press, 1986), 5.

2. Leonard DeGraaf, "Corporate Liberalism and Electric Power System Planning in the 1920s," *Business History Review* 64 (1990): 5; see also William J. Hausman and John L. Neufeld, "The Economics of Electricity Networks and the Evolution of the U.S. Electric Utility Industry, 1882–1935," *Business and Economic History Online* 2 (2004): 1–26.

Pennsylvania Governor Gifford Pinchot and engineer Morris L. Cooke envisioned the potential for an interconnected power system in different terms. Rather than merely serving the needs of the power companies, Pinchot and Cooke's proposed Giant Power Corporation would serve as a mechanism for lowering electric power costs, leading to more efficient farm work, the decentralization of industry, and, ultimately, "a revolution in social and economic relationships." Pinchot, in particular, was an early champion of centralized economic planning in the name of governmental efficiency and conservation of natural resources. Cooke, meanwhile, exemplified the progressive engineer, regarding "professionally trained experts" as the key to "cheaper, more efficient, and more honest government."[3] Herbert Hoover, as secretary of commerce and later as president, held a similar view of the importance of engineers as shapers of policy, but while Hoover believed they should serve the interests of the existing corporate order, Cooke believed engineers should be the vanguard of a challenge to corporate power.[4]

Social scientists, too, would be part of this vanguard. As Thomas P. Hughes explained in *American Genesis*, "electric power was to be the technological agent for the transformation of regions. Progressive politicians would prepare the legislative ground; and social scientists—some spoke of "human engineers"—would preside over planned development."[5] One of the most influential social scientists advocating this agenda was Lewis Mumford, whose vision of the American future focused on the use of electric power and motorized vehicles to bring the benefits of modern life to rural areas while avoiding the concentration of noise, population, and pollution associated with the Industrial Revolution. The region, rather than the metropolis, would be the basic unit of economic development and therefore of social and economic planning as well.[6]

The economic stress of the Great Depression focused new attention on the relationship between the existing corporate order and national interest, and thus also on the potential role of the federal government in managing economic and social resources. One cornerstone of this process was managing the relationship between

---

3. DeGraaf, "Corporate Liberalism," 12–15. See also Thomas P. Hughes, *American Genesis: A Century of Invention and Technological Enthusiasm, 1870–1970*, 303–5.
4. DeGraaf, "Corporate Liberalism," 25–29.
5. Hughes, *American Genesis*, 354.
6. Ibid., 355–60.

production and consumption in a period of "economic maturity." As Franklin Delano Roosevelt put it: "our task now is not discovery or exploitation of natural resources, or necessarily producing more goods. It is the soberer, less dramatic business of administering resources and plants already in hand, of seeking to reestablish foreign markets for our surplus production, of meeting the problem of underconsumption, of adjusting production to consumption."[7] For more than forty years, at least one of Roosevelt's New Deal agencies would embody that commitment to "meeting the problem of underconsumption." The Tennessee Valley Authority, originally conceived by Senator George Norris of Nebraska as a way of repurposing the Army Corps of Engineers' Wilson Dam in Alabama, quickly became a significant element of Roosevelt's vision for a new America. Congress authorized the creation of TVA for several specific purposes, as the preamble to the TVA Act makes clear:

> To improve the navigability and to provide for the flood control of the Tennessee River; to provide for reforestation and the proper use of marginal lands in the Tennessee Valley; to provide for the agricultural and industrial development of said valley; to provide for the national defense by the creation of a corporation for the operation of Government properties at and near Muscle Shoals in the State of Alabama, and for other purposes.[8]

TVA's purpose was manifold—it could be whatever its supporters or detractors imagined. It could be a farm agency, a planning board, a park service, a transportation system, a flood control program, or even a power company. Roosevelt's message to Congress requesting passage of the act described the need thusly:

> a cooperation clothed with the power of Government but possessed of the flexibility and initiative of a private enterprise. It should be charged with the broadest duty of planning for the proper use, conservation and development of the natural resources of the Tennessee River drainage basin and its adjoining territory for the general social

7. Franklin D. Roosevelt, "New Conditions Impose New Requirements upon Government and Those Who Conduct Government," in Roosevelt, *Public Papers and Addresses,* comp. S. I. Rosenman (New York, 1938–1950), 751, quoted in Gregory B. Field, "'Electricity for All': The Electric Home and Farm Authority and the Politics of Mass Consumption, 1932–1935," *Business History Review* 64 (1990): 32-60.
8. The Tennessee Valley Authority Act of 1933, 48 Stat. 58–59, 16 U.S.C. sec. 831.

and economic welfare of the Nation. The Authority should also be clothed with the necessary power to carry these plans into effect. Its duty should be the rehabilitation of the Muscle Shoals development and the coordination of it with the wider plan.[9]

Roosevelt asserted that such a plan, "if envisioned in its entirety, transcends mere power development; it enters the wide fields of flood control, soil erosion, reforestation, elimination from agricultural use of marginal lands, and distribution and diversification of industry."[10] In other words, if successful, it could prove Norris, Mumford, and other advocates of regional planning correct in their claims that integrated development would benefit an entire region.

Despite many claims that the agency's production of electric power was a mere "byproduct" of its broader planning mission, and despite the probably intentional omission of any mention of electric power in the preamble to the TVA Act, a major part of TVA's purpose was to complete the work of rural electrification in the Tennessee Valley, a mission that placed it at cross-purposes with the Tennessee Electric Power Company (TEPCO) and other subsidiaries of Commonwealth and Southern, a vast holding company. Where TEPCO had declined to offer service to rural customers, citing higher infrastructure distribution costs per unit of power consumed, TVA insisted that rural customers be a primary target of its efforts. TVA's first chairman, Arthur E. Morgan, forcefully articulated the agency's mission in terms of rural planning, but it was his rival on the TVA board, David E. Lilienthal, who would translate that mission into aggressive promotion of electric power consumption.[11]

Lilienthal initiated the creation of the Electric Home and Farm Authority (EHFA), a TVA subsidiary aimed at subsidizing and encouraging the sale of electric appliances. Lilienthal's approach has been ably summarized by Gregory B. Field:

9. Franklin D. Roosevelt, "A Suggestion for Legislation to Create the Tennessee Valley Authority," message to Congress dated April 10, 1933, in *The Public Papers and Addresses of Franklin D. Roosevelt with a Special Introduction and Explanatory Notes by President Roosevelt*, vol. 2, *The Year of Crisis, 1933* (New York: Random House, 1938), 122–29.

10. Roosevelt, "A Suggestion for Legislation to Create the Tennessee Valley Authority," 122–29.

11. Hughes, *American Genesis*, 365–76.

[Lilienthal] had to create a much greater market for power across the region so that the TVA could justify its hydroelectric production and more readily acquire territory from the private companies. Facing the need to create demand, Lilienthal came to realize that electric rates were only part of the problem. Even if rates were to come down to acceptable levels, most people did not have load-building appliances that would consume the affordable electricity.[12]

Field argues further that the EHFA represented a "growth-oriented economic agenda" built around government promotion of mass consumption that would not fully take root elsewhere in the Roosevelt administration until the 1940s. Mass consumption would mean a visibly improved standard of living, as measured by access to new home technology; it would also mean increased demand for home appliances and more demand for TVA electricity. Thus, under Lilienthal's patronage the EHFA measured its success in the number of new water heaters, refrigerators, and stoves sold, and the number of kilowatt-hours consumed in turn became a yardstick of both TVA's engineering and marketing success. One of Lilienthal's key advisors in the creation of the EHFA was Giant Power's mastermind, Morris L. Cooke.[13]

After the Supreme Court struck down the National Recovery Act in 1935, thereby significantly undermining FDR's vision for government management of the "mature economy," Congress rechartered the EHFA under significant new limitations and transferred authority for administering it from TVA to the Rural Electrification Administration (REA). Promotion of consumption was a lower priority for REA, though EHFA's activities expanded beyond the Tennessee Valley to a national level. But even after the EHFA disappeared from TVA's mammoth list of responsibilities, TVA would continue to justify its existence in terms of the number of kilowatt-hours consumed.[14]

Of the many New Deal programs, the Tennessee Valley Authority stands out for several reasons—its ambition, the scope of its activities, and certainly the controversies surrounding its creation. The

12. Field, "'Electricity for All,'" 32–60, quoted material from p. 35. Field also notes that the promotion of power consumption was a widespread marketing strategy among privately owned power companies.
13. Field, "'Electricity for All,'" 32–60.
14. Ibid., 55–60.

early years of the agency have also attracted an enormous amount of attention from historians, for mostly those same reasons. A great deal of the historiography of TVA has dealt with the internal struggles within its first board of directors, especially with the conflict between the idealistic and idiosyncratic Arthur E. Morgan and the pragmatic, polished David E. Lilienthal. Most scholars have described Morgan as the more intransigent of the two, though Lilienthal, too, could be difficult at times.

Lilienthal initially viewed TVA's purpose primarily in terms of generating inexpensive electric power. This would encourage economic development, especially industrialization, while also forcing private power companies to acknowledge the economic viability of rural electrification. During his tenure on Wisconsin's Public Service Commission, Lilienthal had developed a deep distrust of the motives of privately owned utilities, and he displayed a bellicose attitude toward such companies. When Morgan and others advocated entering into a pooling arrangement with Commonwealth and Southern, Lilienthal reversed his earlier position and fought against the proposal. Lilienthal saw the private companies as a threat to TVA and viewed any attempt at compromise as dangerous. Morgan asserted that the pool would favor whichever system—private or public—could provide the power most efficiently, and he saw Lilienthal as driven more by political considerations than by the proper engineering criteria. The argument over the pool helped to precipitate a growing feud between the two men, with the third board member, Harcourt Morgan, generally siding with Lilienthal. In the end, the Supreme Court upheld the constitutionality of the TVA Act, Commonwealth and Southern turned over the rights to distribute power in the region to TVA in return for a large sum of money, and TVA received a de facto monopoly on the generation of electric power in its service area.

Lilienthal won both the internal battle with Morgan and the "power fight" with Commonwealth and Southern, and those victories gave him the opportunity to set the tone for TVA's next thirty-plus years. He did so in a landmark treatise entitled *TVA: Democracy on the March*. *Democracy on the March* is a complex apologia for the TVA approach. One of its major themes is what Lilienthal called "decentralized administration of central authority." This seeming paradox is, in Lilienthal's view, a compromise between the need for centralized acquisition of resources (which an economist would call "economy of scale") on the one hand and the need to make sure that those resources are used in a way that provides the greatest benefits for both national and local populations. *Democracy on the March* marks Lilienthal's effort to reconcile the TVA Act's broad endorsement of regional planning with his own view that the agency was, first and foremost, a public utility.

Democracy on the March goes beyond the earlier assumption that the presence of certain appliances marked a comfortable middle-class existence. It does so by celebrating not the mere presence of these appliances, but the presence of *the electric versions* of those appliances: "Electric ranges are used in nearly three fourths of the homes in Nashville which have electricity. Among the more than one million homes served with TVA power it is estimated that 75,000 are heated entirely with electricity. Out of every four new homes being built, three are heated electrically."[15] Lilienthal's rhetoric further asserted with a flourish the moral and technical superiority of electricity over other forms of energy. He continued, "electricity is the most humane and the most efficient form of energy. It is mobility itself: It can be brought to people; people need not be brought to the source of energy. Electricity symbolizes the multiplication of human energies through science."[16] Thus, in Lilienthal's rhetoric, the use of electricity in homes and on farms is a patriotic, moral, and technologically superior activity, and the production of electricity is inherently a democratic priority.

Another theme in *Democracy on the March* is the role played by experts such as engineers, managers, and scientists. In Lilienthal's view, these experts are the best defense against TVA's becoming politicized—by which he meant becoming a tool for political patronage. Lilienthal claimed that the relationship between "the people" and "the experts" was "of the greatest importance in the development of the new democracy. For the people are now helpless without the experts—the technicians and managers." Apparently aware of the potential that this distinction between "the people" and "the experts" might create obstacles, Lilienthal advised that the two groups must cooperate as much as possible; thus, "the experts" should have as much firsthand experience of the needs of

15. David E. Lilienthal, *TVA: Democracy on the March* (New York: Harper & Brothers, 1953), 21.
16. Ibid., 60.

"the people" as possible. One might nonetheless argue—as indeed many TVA critics have—that placing the decision-making processes in the hands of unelected officials reduces the agency's accountability to the people it seeks to serve. Lilienthal's defense is somewhat diffuse and boils down to the agency's popularity among the region's inhabitants.[17]

Thanks to the Second World War, work in the Tennessee Valley was abundant by 1943. Along with the rest of the nation, the valley found its entire infrastructure committed to the war effort, and TVA electricity was central to that transformation. In particular, East Tennessee was heavily involved because of two enterprises: the Aluminum Company of America's large operations in the town of Alcoa and the atomic weapons research taking place at the then-secret site now known as the town of Oak Ridge. Aluminum was essential for making airplanes and for other applications in which weight was critical, and Senator Harry Truman famously complained of the sudden (possibly artificial) shortage of the metal: "We would be willing to buy aluminum from anybody. If we could get it. I don't care whether it is the Aluminum Company of America or whether it is Reynolds or Al Capone."[18] The aluminum shortage was a major factor in the decision to rush construction of several dams in the area, including Cherokee, Douglas, and Fontana. Aluminum smelting and rolling required enormous quantities of electricity, but an even greater demand for power came from about thirty miles northwest of the Alcoa operation. There, at the sprawling "Clinton Engineer Works," later known as the town of Oak Ridge, electric power was used to develop nuclear weapons. By 1945, no longer would Lilienthal have to manufacture demand by pushing residential customers to buy more electric kitchen appliances;

---

17. Ibid., 120–24.
18. Truman made this oft-quoted statement during the testimony of I. W. Wilson, vice president in charge of operations, ALCOA. See *Hearings before a Special Committee Investigating the National Defense Program*, United States Senate, 77th Congress, 1st. Sess. (Washington, DC: Government Printing Office, 1941), pt. 3, 899–922. On the question of whether the aluminum shortage was artificially created by Alcoa, see United States Senate, 77th Congress, 2nd Sess., *Special Committee Investigating the National Defense Program*, Report no. 480, parts 5 and 6, issued January 15 and March 30, 1942, cited in Gregory Field, "From Regional Development to National Defense: TVA, World War II, and the Making of the Military-Industrial Complex," paper presented to the Second Wave/Southern Industrialization Conference, Georgia Tech, June 1998.

indeed, TVA's much-celebrated system of hydroelectric generators would have to be supplemented with the Watts Bar Steam Plant, an utterly conventional coal-fired power plant. Moreover, the impetus for TVA's continued growth would come from industrial rather than residential customers.

Most analysts have assumed that the agency's increased emphasis on electric power and the concomitant decrease in attention to other dimensions of regional development can be attributed to World War II and the Cold War. But historian Gregory Field asserts instead that World War II merely "provided a new and extremely effective justification for Lilienthal's policies, accelerating rather than redefining tendencies already in place at the Authority." Lilienthal had foreseen the war and TVA's role in it and in June 1940 had sent Roosevelt a memorandum calling for a radical expansion of the agency's generating capacity in the name of national defense. Roosevelt agreed that such expansion was necessary but implied that it should not be presented in a way that might seem self-serving on the part of TVA. This leaves aside the fact that, as head of the agency's power programs, Lilienthal stood to gain significant influence as a result; moreover, as a national defense agency, TVA would not be plagued with questions about its value. As Field put it, "Oak Ridge, Alcoa and Monsanto justified the ambitions Lilienthal had for his power program in a way that all the refrigerators in Florence [Alabama] and Tupelo [Mississippi] never could." They would continue to do so in the Cold War as well.[19]

Lilienthal's notion of "democracy on the march" depended heavily on the notion that "the experts" would understand and take appropriate care to safeguard the needs of "the people." Even if we grant Lilienthal's assertion that the decision-making process at TVA took sufficient account of the needs of the local population, which is itself questionable, it is clear that after 1945, with Lilienthal's departure from the agency, TVA began to behave more like a technology-driven bureaucracy and less like Lilienthal's rhetorical "democracy on the march."[20] Erwin C. Hargrove, in his *Prisoners of Myth*, argues that post-Lilienthal members of the TVA board of

19. Field, "From Regional Development to National Defense," 3.
20. Arguably the most influential critic of Lilienthal's claims in this vein is Phillip Selznick, whose *TVA and the Grass Roots* (Berkeley: University of California Press, 1949) deals largely with the conservative bent of the agency's agricultural programs.

directors have been haunted by the agency's early history and have made decisions based not on the needs of their own time but on an outdated conception of the agency's mission. In the 1960s, TVA's chairman, Aubrey J. "Red" Wagner, attempted to revive TVA's commitment to regional development by creating the Office of Tributary Area Development (TAD). The program, in Wagner's view, would accomplish on a smaller scale what he believed the larger projects of the Lilienthal years had done: it would stimulate economic development, involve local residents in the planning process, provide real benefits in resource management, and restore a sense of purpose and unity among the agency's increasingly divergent departments. Few, if any, of these goals were realized by the TAD program, but Wagner pressed on, fervently believing in engineering and planning as the keys not only to the region's continued growth but to defining TVA's continued relevance in any terms beyond the generation of inexpensive electric power to feed the growing military-industrial complex.[21]

Though it is clear in retrospect that the late 1960s saw a dramatic but temporary expansion of the industrial power needs of the Tennessee Valley, TVA planners at the time saw it as the beginning of a new wave of industrialization that would necessitate the construction of an ever-larger number of power plants, and thus the agency embarked on a massive expansion of its power program. This expansion included the experimental Raccoon Mountain "pumped storage" facility, the controversial Tellico Dam project, and the introduction of nuclear power to the valley. All of these would stimulate controversy.

Perhaps the most technically intriguing TVA project of the period is the Raccoon Mountain facility. Described as a "pumped storage" plant, Raccoon Mountain uses off-peak electricity from other plants to pump water up to a mountaintop reservoir. In times of peak demand, the water can be released, and the one thousand–foot drop from the reservoir to the generators located *inside* the mountain allows the plant to provide about sixteen hundred megawatts of electricity, making it TVA's largest hydroelectric plant. Though it may sound like a description for a futuristic movie set, Raccoon Mountain has been a centerpiece of TVA's load management strategy since it

21. Erwin C. Hargrove, *Prisoners of Myth: The Leadership of the Tennessee Valley Authority, 1933–1990* (Princeton: Princeton University Press, 1992), 169–71.

opened in 1978. Since it can only run for twenty-two hours before the reservoir is dry, Raccoon Mountain does not add to the agency's capacity for constant power generation; rather, it exists to improve the system's response to maximum load conditions. Typical summer operation involves using nighttime power from other plants to refill the reservoir, then allowing the reservoir to drain through the generators during the hot late-afternoon peak-usage period.[22]

The construction of the Raccoon Mountain plant marked a turning point for TVA, as it signified a new approach to hydroelectric power, one that went beyond conventional dam building. Similarly, the proposed Tellico Dam included no additional generating units but was envisioned as a helper dam for the larger Fort Loudoun dam on the Tennessee River. Both Raccoon Mountain and Tellico were aimed at improving the efficiency of the existing system rather than operating as independent hydroelectric sites. Indeed, since there were no viable sites left in the valley for additional large-scale hydroelectric dams, TVA planners analyzed the cost of the Tellico project by comparing it to the cost of producing an equivalent amount of nuclear power.[23] The Tellico project has already been examined extensively by historians, but several points bear mentioning here. First, the local populace largely opposed the project in the beginning. Second, a large environmentalist lobby did so as well, for overlapping but distinct reasons. Third, TVA resisted calls for an environmental impact study, claiming that the National Environmental Policy Act of 1969 could not be applied to a federal project whose construction was already under way. Fourth, the vision for the Tellico area's redevelopment—the ambitious Timberlake community—reflected then-current thinking on town planning and would have required massive private investment, which was initially anticipated from the Boeing Aerospace company. Fifth, once the construction was largely complete, though the environmentalists kept complaining, local sentiment generally supported reservoir development rather than allowing the destruction of the dam and the return of the free-flowing river.[24]

---

22. *http://www.tva.gov/sites/raccoonmt.htm*, viewed on April 18, 2008; North Callahan, *TVA: Bridge over Troubled Waters* (Cranbury, NJ: A. S. Barnes & Co., 1980), 297–98.

23. Tennessee Valley Authority, *Alternatives for Completing the Tellico Project* (internal report, December 1978), 104–6.

24. Ibid., 1–43.

The gains in system efficiency afforded by Raccoon Mountain and Tellico would, at best, serve as stopgaps. By the mid-1960s, before either Raccoon Mountain or Tellico could be completed, TVA's engineers had concluded that nuclear power offered the best opportunity to maintain TVA's historically low rate schedule in the face of presumed massive growth in demand. Coal prices had risen to the point that additional steam plants were no longer likely to be cost-effective. So the agency embarked on an ambitious nuclear plant construction plan. But while they could produce electric power very cheaply once built, the nuclear plants themselves were far more expensive to construct and required considerably longer to build than the steam plants. In addition, safety standards for the plants were often a moving target, leading to greater delays and cost overruns than in previous TVA projects.[25]

All three developments—Raccoon Mountain, Tellico, and the nuclear program—were aimed at resolving an anticipated shortage of electric power in the Tennessee Valley. This anticipated shortage was given additional immediacy by the experience of the great blackout that darkened much of the northeastern United States in 1965. It is possible that the decision to expand TVA's generating capacity was made under unrealistic assumptions. One such assumption is that the growth rates seen in the agency's first thirty years would continue indefinitely; another is that such growth could be fueled indefinitely on a steady supply of inexpensive electricity. The net result of these assumptions is that the power program overshadowed virtually all other TVA operations, from navigation and flood control to resource development and fertilizer programs. As one observer pointed out in 1969, "although official statements continue to accord higher priority to broader goals of regional resource development, social objectives, and the like, realistically the tail now wags the dog."[26]

One thing stands out in annual reports from TVA's middle period, the 1950s through the late 1970s. Regardless of who held the chairmanship of the agency, TVA continued to justify its mission—and trumpet its successes—in terms of a steady increase in the consumption of electric power. Even more than production, *consumption*

25. Callahan, *Bridge over Troubled Waters*, 311–15.
26. Victor C. Hobday, *Sparks at the Grassroots: Municipal Distribution of TVA Electricity in Tennessee* (Knoxville: University of Tennessee Press, 1969), 17.

was deemed evidence of TVA's continuing relevance, because consumption was evidence of demand. The approach of celebrating consumption ran counter to the prevailing political economy of the 1970s, which was marked by scarcity of energy—especially the shortage of fossil fuels triggered by the 1973 Arab oil embargo. In the late 1970s excessive consumption of energy was politically unpopular, and TVA found it no longer convenient to justify its existence through the success of its consumption programs. Instead the agency had to shift focus away from promoting consumption toward promoting conservation—an ironic position given the agency's previous forty years. While TVA had intermittently provided guidance to residential and small business customers seeking to reduce their energy costs, a skeptic might compare these earlier conservation efforts to liquor ads that advise drinkers to "enjoy responsibly." Nonetheless, by the late 1970s, TVA had begun to take conservation seriously, as costs spiraled out of control and TVA power could no longer be guaranteed to be cheaper than that provided by private companies in adjoining regions.

The man responsible for overseeing the shift was S. David Freeman, appointed chairman during the Carter presidency. Freeman called for the agency to return to the spirit of its founding decade. As he told the *New York Times* in a 1978 interview, "We need to go back to the Rooseveltian vision of TVA as a great innovator, a place where things are tested out and proven—a living laboratory and the first place where national energy policy becomes a reality."[27] In his bid to succeed Wagner, Freeman wrote to President Carter that the agency should be involved in promoting conservation through education, changes in the rate schedules, and alternative energy sources. Freeman's TVA would be more responsive to the public and more innovative in its approach to energy policy; Freeman's barrage of notes to general manager Lynn Seeber and other officials included demands for "solar co-ops, a bio-mass program, wood alcohol to replace gasoline, railroad electrification" and an electric vehicle program. Furthermore, under Freeman the agency would abandon the long-standing paradigm of measuring its success in terms of increased consumption, as both national policy and regional needs now focused more attention on the need for conservation of electric power. Though Freeman was ostensibly chosen because of his

27. Quoted in Hargrove, *Prisoners of Myth*, 201.

energy expertise, his populist political rhetoric together with his credentials as an environmentalist certainly fit both the tenor of the times and the Carter administration's priorities. With Freeman at the helm, TVA's intransigence in complying with Environmental Protection Agency guidelines would also swiftly come to an end.[28]

The spiraling cost of constructing nuclear power plants in the face of economic stagnation brought TVA to a crisis point in the early 1980s. How serious was this crisis? Serious enough that the agency's manager of power, Hugh G. Parris, addressed the Union-Management Cooperative Committee with a speech provocatively entitled "The New TVA: Managing in a No-Growth Environment." Parris warned that cutbacks in future projects and in current staff were inevitable because the economic parameters under which the agency operated had changed:

> It is possible for the market for any product to change drastically. . . . Sometimes it is not possible to "produce your way out of a problem." I know many of us naturally think we can and should go back to promoting the sale of electricity as a principal way of getting out of this tight spot. But that will not work if the price is not competitive, and the market doesn't want that much of your product anymore. When we sold electricity so hard and successfully before, it was low in cost, plentiful, every increment of new generation was lower in cost, and people wanted it.[29]

Chairman David Freeman reached the conclusion that radical cutbacks in the agency's construction plans were the only solution to the cost problem. Accordingly, the agency's original plan for seventeen nuclear reactors at eight sites was scaled back to a total of nine reactors by 1984, of which five were complete at Browns Ferry and Sequoyah, with four still under construction at Bellefonte and Watts Bar. Safety problems eventually forced additional closures.[30]

Under these circumstances, the only viable direction TVA could take was to move toward conservation and away from new construction. A TVA pamphlet from 1983 proclaimed the agency's

28. Ibid., 195–201; quoted matter from 200.
29. Hugh G. Parris, "The New TVA: Managing in a No-Growth Environment," remarks to the TVA-TVTLC Union-Management Cooperative Committees, 36th Annual Valley-Wide Meeting, Nashville, Tennessee, July 29, 1982 (Knoxville: TVA Information Office, 1982).
30. Hargrove, *Prisoners of Myth,* 244–55.

efforts in terms that would have made Lilienthal's generation blanch: "Energy Conservation is, in a real sense, a source of 'new energy supply.' The more energy saved, the less need for expensive new plants to provide that energy and less drain on the pocketbooks of electricity consumers. TVA-sponsored conservation measures are already estimated to be saving enough energy to offset the need for an additional multibillion-dollar nuclear generating plant." The pamphlet went on to boast that the agency was sponsoring the installation of solar water heaters and wood stoves, as well as working with modular home manufacturers to develop and demonstrate solar home designs and researching solid waste cogeneration and methanol fuels—all measures aimed largely at residential customers. Indeed, in late 1978 the agency offered 120 wood stoves to residents of six low-income counties in East Tennessee, and the program was so successful—heating bills dropped by 60 to 80 percent—that it was later expanded to another one thousand customers in north Georgia.[31]

Yet even as it was promoting overall energy conservation and exploring alternatives to electric power for certain applications, TVA still advocated the use of electric power as an alternative to fossil fuels. Because of their limited range on existing battery technology, electric vehicles even today are suitable primarily for applications in which the distance traveled in a given day is small; thus, they are better suited to government or industrial fleet use than for most private users. TVA's fascination with electric automobiles first became evident in 1961, when the agency purchased a Henney Kilowatt automobile for testing and demonstration purposes. The Henney Kilowatt, a Renault Dauphine converted to electric propulsion by a subsidiary of the Eureka Vacuum Cleaner Company, had a top speed of only thirty-five miles per hour and a range of only fifty miles on a charge—yet the agency kept it in service for eight years before giving up. Only 120 Henney Kilowatts were sold, and 30 of those went to electric utilities. Eight years after mothballing the agency's only Henney Kilowatt, TVA tried again, purchasing six Electra Van 500s. These vehicles, converted from Subaru microvans not sold in the U.S., proved little more than a curiosity, though four of them remained in use on TVA property as late as 1981. Though

31. Tennessee Valley Authority, "TVA and Electric Power" (Knoxville: TVA Information Office, 1983); Callahan, *Bridge over Troubled Waters*, 365.

the Electra Vans might seem laughable—they were little larger than typical golf carts—they did signal a new interest on the part of the agency. The following year, the agency worked with the Electric Power Research Institute to demonstrate the viability of electric cars on a larger scale, this time using converted Volkswagen buses, and in 1979 a still-larger effort coordinated by the Department of Energy gave TVA yet another opportunity to display its commitment to shifting consumption from fossil fuels to electric power, ostensibly in the name of overall environmental conservation. Despite Chairman David Freeman's insistence that "we are going to show [Detroit automakers] that the Tennessee Valley can lead in making the use of electric cars and electric trains possible," TVA's actual commitment to the electric vehicle program was modest at best and likely served more political than technical purposes.[32]

TVA's promotion of certain technologies was sometimes driven less by regional needs than by the agency's continuing quest for political, social, and economic relevance. The words Elliot Roberts wrote of the agency back in 1955 could still hold true thirty years later for the power program, if not necessarily for the whole agency: "the pattern . . . which emerges from these events is not one of partnership, nor of continuing interaction between levels of government in working out joint policy, nor of shared responsibility. It is a pattern of firm federal control of basic programs under accountability to the Executive and the Congress, with intermittent attempts to make common cause with state agencies concerning fringe issues."[33] But Roberts went on to note: "The TVA has been free to execute its tasks alone. It has been able to make decisions rapidly, to take action on a basin-wide scale, and to deal with the river in terms of coordinated engineering and managerial judgments, unhampered by the need to renegotiate with the states the basic policies

---

32. Tennessee Valley Authority, "Tennessee Valley Authority Electric Vehicle Program—Electric Vehicle Test Facility—Chattanooga, Tennessee" (Knoxville: TVA Information Office, 1981); "Brief History of the Henney Kilowatt," *http://www. ccds.charlotte.nc.us/~jarrett/EV/history.php*, viewed on April 18, 2008; *http://member. newsguy.com/~apeweek/ElectraVan.html*, viewed on November 30, 2006; Hargrove, *Prisoners of Myth*, 221–23. A full-scale examination of TVA efforts to promote industrial conservation is beyond the scope of this essay.

33. Elliot Roberts, *One River—Seven States: TVA-State Relations in the Development of the Tennessee River* (Bureau of Public Administration, University of Tennessee: University of Tennessee Record Extension Series, vol. 31, no. 1 (June 1955): 90. See also Hobday, *Sparks at the Grass Roots*, 30–31ff.

laid down by Congress."[34] That may have been true in 1955, but by the 1980s, it was clearly no longer the case. The long stalemate between TVA and the Environmental Protection Agency, the largely futile attempts at cooperation between TVA and the Department of Energy on the subject of alternative energy sources, and the crisis of the agency's nuclear programs all testify to the growing complexity and difficulty of defining the agency's role in a changing region and a changing nation.

Rarely did TVA actually "invent" anything in the sense of devising an entirely new solution to a physical, electrical, mechanical, or chemical problem. Yet technology has been central to the agency's mission since its founding. TVA has served as a promoter of electric consumption through increased availability of electric appliances. It has played a supporting role in the development of new fertilizer applications and new weaponry. It has served as a testing ground for new environmental programs, whether through the use of "scrubbers" on its coal-fired plants or the use of electric vehicles as an alternative to the internal combustion engine. It has served as a mechanism through which technology developed by others could be efficiently demonstrated to potential users. In all cases, though, the agency has displayed a firm belief that technology, properly developed and applied, possesses the capacity to improve the quality of life for people in the Tennessee Valley and the rest of the nation.

34. Roberts, *One River—Seven States*, 94.

Telemedicine

An Important Component in
Arizona's Economic and Social Development

YONEYUKI SUGITA

THIS ESSAY LOOKS AT SOME OF THE FACTORS THAT UNDERPIN Arizona's recent emergence as a star performer in the Sunbelt and explains Arizona's limitations, focusing on telemedicine as a case study.[1] Telemedicine embodies many of the characteristics of Arizona's strategy for development in the late twentieth century. Attaining statehood in 1912, Arizona did not fight the Civil War and was rarely included in the "South" in any traditional sense; however, between Reconstruction and the early twentieth century, it was mostly southerners who migrated to Arizona Territory and engaged in agriculture and mining, and the territory became, in a way, an extension of the traditional South.

During World War II, Washington poured a large amount of money into what we now think of as the Sunbelt, including Arizona, in order to establish military bases. These bases contributed to the emergence of defense industries in this region. This initial push by the federal government precipitated the development of the Sunbelt.

The traditional concept of the South carries negative connotations of racial discrimination, rebellion against the North and secession from the Union, and backward rural areas based on an agricultural economy. The concept of the modern Sunbelt brings to

---

1. There is no consensus on which states constitute the Sunbelt; however, most people agree that the following states are included: Arizona, California, Florida, Louisiana, Georgia, Nevada, New Mexico, and Texas. Many people would add South Carolina, Mississippi, Arkansas, and Alabama, and North Carolina, Virginia, and Tennessee may sometimes be considered Sunbelt states.

mind bright images: much sunshine, many capital- and knowledge-intensive industries, and a highly educated white-collar population. Arizona, originally an extension of the traditional South, has transformed itself into a member of the Sunbelt; however, in reality, the social and economic fabric beneath the surface remains similar to that of the traditional South.

Economic and political power has been shifting from the Frost-belt to the Sunbelt since the late 1960s. The Sunbelt offers not only economic opportunities but also good quality of life, including advanced and efficient health-care services. Because the Sunbelt states came late to industrialization, they tended to adopt a catch-up strategy based on a government-industry-university collaboration. During and after World War II, initial industrial development in the Sunbelt was led and promoted by the region's three core states: California, Texas, and Florida. But more recently, neighboring Sunbelt states such as Nevada, Arizona, and North Carolina have been experiencing higher economic growth. Arizona has been the star performer.

The Development of the Sunbelt

From the late nineteenth century to the end of World War II, the southern states were considered the Solid South by the Democratic Party, providing party candidates with reliable voter support year after year, especially in presidential elections. The Solid South began to crumble when the Democratic administration of President Harry S. Truman demonstrated support for the civil rights movement. In the 1960s, Presidents John F. Kennedy and Lyndon B. Johnson, both Democrats, actively promoted the civil rights movement, culminating in the passage of the Civil Rights Act of 1964. This was a watershed in African Americans' struggle for fairer treatment in American society. This support by the Democratic Party for increased civil rights protection for African Americans disillusioned white American conservatives and those opposed to eradicating racial injustice in the Democratic Party. In consequence, the conservative wing of the Republican Party adopted a southern strategy to take advantage of this disillusionment, beginning with the 1968 presidential campaign. The more conservative Republican Party tried to woo disenchanted whites by backing states' rights and extolling respect for law and order in American society. To many southern whites, this strategy demonstrated the party's willingness to reach out for their votes and close its eyes to discrimination against African Americans in the southern states.[2]

The new Republican strategy appealed to whites who resented both the progress in civil rights and the various grassroots campaigns for desegregation under way throughout the United States. The party sought to fuse conservative white voters to its traditional business constituency in order to pave the way for an expanded Republican constituency that could dominate southern politics. In the 1970s, the term *Sunbelt* was starting to come into use—the word was coined in 1969 by Kevin P. Phillips, a Republican adviser to and political strategist for President Richard Nixon.[3] In Phillips's view, politics in these warm and sunny regions of the United States were changing, transforming the Sunbelt into the new and conservative bedrock of the Republican Party.

When the Republican Party took control of the White House following Nixon's victory in 1968, it began to channel economic resources to states in the Sunbelt in a bid to secure voter loyalty among southern whites. The Departments of Defense, Health and Human Services, Transportation, Agriculture, the National Science Foundation, the National Aeronautics and Space Administration (NASA), and other federal agencies devoted massive funding to all kinds of public projects. Federal contracts, direct and indirect subsidies, research and development funds, federal grants, financial guarantees, and other forms of assistance were directed to the region.[4] In regard to spending on defense, this form of largesse had already been making a critical contribution to economic growth in the Sunbelt since World War II. In fact, defense spending was credited with helping to create a new South, mainly because the

2. For the development of the Sunbelt, see Carl Abbott, *The New Urban America: Growth and Politics in Sunbelt Cities* (Chapel Hill: University of North Carolina Press, 1981); Richard M. Bernard and Bradley R. Rice, eds., *Sunbelt Cities: Politics and Growth since World War II* (Austin: University of Texas Press, 1983); Raymond A. Mohl, ed., *Searching for the Sunbelt: Historical Perspectives on a Region* (Knoxville: University of Tennessee Press, 1990).

3. Kevin P. Phillips, *The Emerging Republican Majority* (New Rochelle, NY: Arlington House, 1969), 139.

4. Tom R. Res, "Arizona Average in Receipt of Federal R&D Funds," *AZB, Arizona Business* (July 2001): 10; Bill Luker Jr., "The Public Sector and Sunbelt Development," *Challenge* vol. 40, no. 4 (July/August 1997): 60–61.

Table 1.
Fastest-Growing Major Cities (Top 10)

RANK	CITY	2002 POPULATION	1990 POPULATION	% CHANGE
1	Henderson, NV	206,153	65,109	216.6
2	Chandler, AZ	202,016	90,703	122.7
3	Las Vegas, NV	508,604	259,834	95.7
4	Plano, TX	238,091	128,507	85.3
5	Scottsdale, AZ	215,779	130,086	65.9
6	Glendale, AZ	230,564	150,867	52.8
7	Mesa, AZ	426,841	290,212	47.1
8	Bakersfield, CA	260,969	183,959	41.9
9	Raleigh, NC	306,944	220,425	39.3
10	Phoenix, AZ	1,371,960	988,983	38.7

Cities with populations of 200,000 or more, based on 2002 U.S. Census Bureau estimates
The World Almanac and Book of Facts 2005, p. 413.

region's warm climate attracted a large concentration of military base sites and training facilities. As a result, the Sunbelt became the U.S. region most dependent on military spending.[5]

The Sunbelt also benefited from federal investment in huge public works projects for building highway networks and from government programs designed to stimulate increased home ownership through lower taxes and mortgage rates.[6] The flow of federal aid attracted both people and capital to the Sunbelt.

The Sunbelt has experienced the most vigorous and rapid growth in economic and political power, and the most significant infrastructural development of all U.S. regions over the past five decades.[7] As table 1 shows, between 1990 and 2002, the majority of the ten fastest-growing major U.S. cities were in the Sunbelt, including five in Arizona alone.

5. Becky M. Nicolaides, "Suburbia and the Sunbelt," *Magazine of History* vol. 18, no. 1 (October 2003); Carl Abbott, "Urbanizing the Sunbelt," *Magazine of History* vol. 18, no. 1 (October 2003).
6. Marc Fink, "Toward a Sunbelt Urban Design Manifesto," *Journal of the American Planning Association* vol. 59, no. 3 (summer 1993).
7. Kirkpatrick Sale, *Power Shift: The Rise of the Southern Rim and Its Challenge to the Eastern Establishment* (New York: Random House, 1975); Carl Abbott, *The New Urban America: Growth and Politics in Sunbelt Cities* (Chapel Hill: University of North Carolina Press, 1987); Bruce J. Schulman, *From Cotton Belt to Sunbelt: Federal Policy, Economic Development, and the Transformation of the South, 1938–1980* (Durham: Duke University Press, 1994); James C. Cobb, *The Selling of the South: The Southern Crusade for Industrial Development 1936–1990* (Urbana: University of Illinois Press, 1993); Luis

Table 2.
Production of Crude Oil by State, 2004 (Thousand Barrels)

RANK	STATE	TOTAL	DAILY AVERAGE
*1	Texas	392,867	1,073
2	Alaska	332,465	908
*3	California	240,206	656
*4	Louisiana	83,411	228
*5	New Mexico	64,236	176

U.S. Energy Information Administration http://www.eia.doe.gov/neic/rankings/
crudebystate.htm
* Sunbelt states

At first, the Sunbelt primarily attracted service industries, but in more recent years a process of full-blown industrialization has taken over, spurred by the rapid growth of businesses in the high-tech sector such as the telecommunications and computer industries.[8] The region has benefited from a number of features that have attracted inward investment of capital: state corporate income tax at lower rates than those in northern states, lower wages, lower rates of union membership, and lower welfare payments.[9] The Sunbelt is also blessed with abundant energy resources such as oil.[10] As table 2 demonstrates, four of the top five crude-oil producing states in the United States are located in the Sunbelt.

As mentioned above, the Sunbelt region also has an attractive climate. Winters are relatively mild, while greater use of air conditioners, especially after World War II, enables residents in the region to

Suarez-Villa, "Regional Inversion in the United States: The Institutional Context for the Rise of the Sunbelt Since the 1940s," *Tijdschrift voor Economische en Sociale Geografie* vol. 93, no. 4 (2002); Bernard L. Weinstein, Harold T. Gross, and John Rees, *Regional Growth and Decline in the United States* (New York: Praeger, 1985).

8. Dirk Hanson, *The New Alchemists: Silicon Valley and the Microelectronics Revolution* (Boston: Little Brown, 1982); Michael I. Luger and Harvey A. Goldstein, *Technology in the Garden: Research Parks and Regional Economic Development* (Chapel Hill: University of North Carolina Press, 1991); Allen J. Scott, *Technopolis: High-Technology Industry and Regional Development in Southern California* (Berkeley: University of California Press, 1993).

9. Cobb, *The Selling of the South*, 262–63; Bob Hall and Bob Williams, "Case Study: Who's Getting Rich in the New South," *Southern Exposure* vol. 6, no. 3 (fall 1978): 95; Larry Sawers and William K. Tabb, eds., *Sunbelt/Snowbelt: Urban Development and Regional Restructuring* (New York: Oxford University Press, 1984), 6–13.

10. Ryo Kawade, "Beikoku Nanbu Keizai no Doko" ["Development of the U.S. Southern Economy"], *Keizai to Gaiko* 714 (November 1981).

get through the hot summer months. In 1960, one-fifth of residential homes in the Sunbelt states were air-conditioned. Two decades later, almost four-fifths had air conditioners.[11]

From an economic standpoint, the Sunbelt became more attractive in the 1970s. Various factors contributed to this development. One such factor was that U.S. hegemony over the international economy began to decline in the late 1960s, partly because of a slowdown of the U.S. domestic economy induced by lavish spending on the Vietnam War. In addition, the oil crisis of 1973 drastically increased manufacturing costs, while Japan's aggressive drive to export to the U.S. market compelled American corporations to reduce product prices. In a bid to maintain profitability, U.S. companies began shifting their operations to the Sunbelt, where energy, labor, and living costs were lower.

Aside from cost considerations, a change in the structure of U.S. industry factors into explanations of why U.S. corporations moved to the Sunbelt in the 1970s. Labor-intensive heavy industries of the postwar era such as iron, steel, and automobiles that were centered in the Northeast and Midwest (the so-called Frostbelt) began to suffer a decline in economic importance and profitability, while newer high-tech and capital-intensive industries such as electronics, telecommunications, and computers took on a more prominent role. In the Sunbelt, these newer industries were spin-offs from the large number of existing defense-related industries.

Sunbelt cities are generally suburban by design, quite different from the cities in the Northeast and Midwest with their distinct urban features. These suburban cities are characterized by decentralization, low population density, and a predominance of single-family housing.[12] Attractive because of these economic opportunities and comfortable living conditions, the Sunbelt has seen a large population influx. In general, those moving to the region were likely to be young and have above-average or high levels of education. When young people graduate from universities in the North, they often move to the Sunbelt for employment. This interregional migration has correlated well with the shifting focus of the U.S. economy from the Frostbelt to the Sunbelt.[13] The growth of the Sunbelt's popula-

11. John B. Boles, *The South through Time: A History of an American Region* (Upper Saddle River, NJ: Prentice Hall, 1999), 2:501.
12. Fink, "Toward a Sunbelt Urban Design Manifesto."
13. Rones, "Moving to the Sun," 12, 16–17.

Table 3.
Top 10 Cities in the U.S. by Population and Rank (2004)
http://www.infoplease.com/ipa/A0763098.html

RANK	CITY	POPULATION
1	New York City, New York	8,143,197
2	Los Angeles, California	3,844,829
3	Chicago, Illinois	2,842,518
4	Houston, Texas	2,016,582
5	Philadelphia, Pennsylvania	1,463,281
6	Phoenix, Arizona	1,461,575
7	San Antonio, Texas	1,256,509
8	San Diego, California	1,255,540
9	Dallas, Texas	1,213,825
10	San Jose, California	912,332

Source: U.S. Census Bureau

tion continues. The U.S. Census Bureau's population estimates show that in 2004, six cities in the Sunbelt were included among the ten largest cities based on population: Los Angeles, Houston, Phoenix, San Diego, Dallas, and San Antonio (see table 3).

Because of the large, rapid influx of people to the Sunbelt, which was mostly underdeveloped until the 1970s, a new regionwide community came into existence. An extensive network of infrastructure, demanding vast public investment, was required to support this community. Investment has come primarily from the state government–directed transportation sector and a mix of public-private spending on projects in the construction, development, and housing sectors.[14] As a result of the significant public spending, the Sunbelt has managed to build up a mature stock of infrastructure, which of course reinforces development in the region.[15]

When industrial development took hold in the Sunbelt, it was initially led by the region's three largest states: California, Texas, and Florida. More recently, neighboring southern states have been experiencing higher rates of growth. These so-called New Sunbelt states, including Nevada, Arizona, Colorado, Utah, Georgia, and North Carolina, were the fastest-growing U.S. states in the 1990s

14. Luker, "The Public Sector and Sunbelt Development."
15. Atsushi Fujioka, "Gasshukoku Nanbu no 'Sunbelt'ka no Keizaiteki Imi," (Jo) ["Economic Significance of Becoming the 'Sunbelt' of the South in the United States" (Vol. 1)], *Ritsumeikan Keizaigaku* vol. 31, no. 3 (August 1982): 131.

with Arizona standing out as particularly successful.[16] The capital of Arizona, Phoenix, once a small rural community in the center of the state, has become the largest metropolitan area in the Southwestern United States. World War II was responsible for the sudden and rapid development of Phoenix. With a climate conducive to military flight training, an inland location that enjoyed an excellent transportation network, and the more recent invention of air conditioning, large military air bases and a number of important defense industries sprang up around Phoenix in the 1940s.[17] Tucson is Arizona's second largest city. The two cities have been expanding their geographical perimeters quite rapidly. Just 17 square miles in 1950, by 1990, Phoenix had grown to 420 square miles. Over the same period, Tucson expanded from 10 to 156 square miles.[18]

After Nevada, Arizona has recently had the second fastest population growth rate of all U.S. states. Its population grew by 40 percent from 3.7 million in 1990 to 5.1 million in 2000. Net migration into the state accounted for nearly 75 percent of this population increase. Most of these new Arizonans were working-age young adults who came in search of employment.[19] Over this same ten-year period, young people in their twenties and thirties accounted for 20.7 percent and 16.1 percent of total immigration into the state.[20] According to the Census Bureau, Arizona's population increased by 20 percent from 2000 to 2006, to more than 6.1 million people, and this population growth is expected to continue.[21] Phoenix, with a population of 1.42 million in 2004, was one of the fastest-growing cities in the United States, ranking sixth in population terms. California is still an attractive location for working, education, and living, but the rising cost of living there has encouraged an exodus from California to neighboring states. Average house prices in Phoenix are less than half those in the San Francisco Bay area. In 2005, approximately 120,000 people moved from California

16. William H Frey, "Shifts in Political Power," *World & I* vol. 16, no. 5 (May 2001).

17. Bradford Luckingham, *Phoenix: The History of a Southwestern Metropolis* (Tucson: University of Arizona Press, 1989).

18. Fink, "Toward a Sunbelt Urban Design Manifesto."

19. Tom Rex, "Arizona Statewide Economic Study 2002: Arizona Economic Base Study and Special Reports" (July 2002).

20. Tom Rex, "Arizona Statewide Economic Study 2002: Demographic Analysis" (July 2002). *http://www.azcommerce.com/pdf/prop/sesreports/Demographics.pdf*

21. U.S. Census Bureau, State and County QuickFacts: Arizona *http:// quickfacts.census.gov/qfd/states/04000.html* (accessed February 25, 2008).

to Arizona, a figure that equates to approximately 2.5 percent of Arizona's total population.[22]

Arizona has a variety of labor and taxation policies that corporations find appealing. For example, Arizona is a right-to-work state. The state imposes no corporate franchise tax, no business inventory tax, no income tax on dividends from out-of-state subsidiaries, and no worldwide unitary tax. Virtually all private business services are exempt from sales tax, and 100 percent of NOLs (normal operating losses) may be carried forward for five subsequent years.[23]

Arizona's high-quality health-care services, low taxes, and affordable cost of living have attracted such world-renowned corporations as Motorola, Intel, Novell, Cisco, America Online, American Express, General Electric, Hewlett-Packard, IBM, Ford, Nissan, Toyota, Honeywell, Avnet, Phelps Dodge, America West Holdings, PetSmart, Dial Corporation, Del Webb, and Insight.[24] In Arizona, half of all employees in the manufacturing sector work for high-tech corporations. Intel, the largest semiconductor manufacturer in the United States, is one example of the many giant U.S. corporations that have committed to placing major operations in the state. In January 2000, Intel announced that it would invest $2 billion in the construction of its first-ever state-of-the-art, high-volume 300mm wafer production facility in Chandler, Arizona. In 2002, the Arizona Department of Administration adopted the latest Novell net business solutions system, Novell NetWare 6, to enable public officials to freely access government information from any location, thus providing state residents with a better service.[25]

Future prosperity and growth in Arizona depend on continuous investment from high-tech, knowledge-intensive industries.[26] In November 1999, then-governor Jane Dee Hull appointed a thirty-six–member steering committee called the Arizona Partnership for the New Economy (APNE) and tasked it with making recommendations on the best way Arizona could adapt to the new economy.

22. "United States: Dreams in the Desert; The South-western Economy," *Economist* vol. 377, no. 8454 (November 26, 2005).

23. "Arizona's Competitive Business Operating Environment," Arizona Department of Commerce.

24. "Arizona's High Quality of Life," Arizona Department of Commerce.

25. Novell press release (April 17, 2002). *http://www.novell.com/news/press/archive/2002/04/pr02035.html*

26. "The High Technology Industries of Arizona: Leveraging Assets to Grow Innovation Economy," Arizona Department of Commerce (February 2001).

APNE strongly recommended that to "attract, retain and grow the type of talent and cutting-edge businesses that fuel success in the new economy, [the state government of] Arizona should support and build communities with vibrant economies and a high quality of life."[27] In 2006, Governor Janet Napolitano, an advocate of pro-growth policies for the state, took the initiative of slashing the business-property tax and providing corporations with tax credits for research and development.[28] On top of creating a probusiness climate, Arizona strongly promotes a government-industry-university development strategy.

Arizona Telemedicine Program (ATP)

The Arizona Telemedicine Program (ATP) is emblematic of Arizona's successful development. It is a high-tech form of long-distance medical care designed to reach into the rural areas of the state. In the field of advanced medical practice, ATP embodies yet another catch-up strategy that places government-industry-university collaboration at the service of a strong state-led initiative to promote economic growth and good-quality lifestyles, including high income and inexpensive living standards.

In the United States, more than 63 million people live in rural communities in an area that covers 80 percent of the country. Because rural inhabitants sometimes live great distances from major medical centers, they have often suffered from lack of access to primary care physicians, let alone to specialists.[29] As of September 2006, fifty-eight areas in the United States were declared "Primary Care Health Professional Shortage Areas," a designation devised by the U.S. secretary of health and human services to identify an area or population experiencing a shortage of primary health-care providers.[30]

27. Arizona Partnership for the New Economy, "An Economy That Works for Everyone: Final Report" (January 2001).

28. "United States: Dreams in the Desert."

29. Teri Randall, "Rural Health Care Faces Reform Too: Providers Sow Seeds for Better Future," *Journal of the American Medical Association* vol. 270, no. 4 (July 28, 1993): 419; Tom Dorr, "Review of the Department of Agriculture's Distance Learning and Telemedicine Program," FDCH Congressional Testimony, June 25, 2003.

30. Arizona Department of Health Services, Division of Public Health Services. *http://www.azdhs.gov/hsd/hpsa.htm* (accessed February 25, 2008).

Arizona, the sixth largest state in the United States, is also one of the most rural. Thirteen of Arizona's fifteen counties are designated rural. Arizona extends over more than 113,900 square miles. The state's rural areas comprise more than 95,000 square miles and host a rural population of about 300,000. According to the Rural Health Office at the University of Arizona, the state suffers from an uneven distribution of health-care resources. Only Tucson and Phoenix have fully modern medical services; those living in rural areas have only limited access to basic medical-care facilities and must spend hours driving to Tucson or Phoenix to receive specialist care.[31]

To improve the quality of medical care for rural residents, Arizona introduced telemedicine. The Arizona legislature defines telemedicine as "the use of computers, video imaging, fiber optics, and telecommunications for the diagnosis and treatment of persons in rural, geographically isolated communities and state institutions."[32] But Arizona is far from being the first to use telemedicine. In the early 1960s, the National Aeronautics and Space Administration (NASA) was using telemedicine technology. In 1983, Massachusetts General Hospital in Boston was the first hospital in the nation to use telemedicine. Although advanced technology holds out great promise for the medical profession, its adoption in the medical field initially faced many formidable obstacles.

Prior to the 1990s, telecommunication technologies were too unstable to be used reliably in health-care services. Many physicians had doubts about their efficiency and complained that the telecommunications devices then available for telemedicine were complicated and wasted too much time.[33] It was recognized that high-resolution video-imaging transmission technology and fast broadband Internet connections were crucial to the effective implementation of telemedicine strategies, but prior to the 1990s the initial cost of installing the necessary information technology (IT)

31. Kevin M. McNeill, Ronald S. Weinstein, and Michael J. Holcomb, "Arizona Telemedicine Program: Implementing a Statewide Health Care Network," *Journal of the American Medical Informatics Association* vol. 5, no. 5 (September/October 1998): 441–42.

32. "Marconi Helps Arizona to Build a Thriving Telemedicine Program." *http://www.marconi.com/media/AZTelemedicine_cs.pdf.*

33. Ibid.; R. Roine, A. Ohinmaa, and D. Hailey, "Assessing Telemedicine: A Systematic Review of the Literature," *Canadian Medical Association Journal* 165 (2001); W. R. Hersh et al., "Clinical Outcomes Resulting from Telemedicine Interventions: A Systematic Review," *BMC Medical Informatics and Decision Making* vol. 1, no. 5 (2001).

equipment was in the neighborhood of $100,000 to $120,000 in a rural clinic. That sum, combined with telemedicine operational costs, proved daunting to many hospitals. In addition, because IT evolves rapidly, constant upgrading of equipment and software might be necessary. Rural areas usually suffer from a poor telecommunications infrastructure and a lack of broadband network services.[34] But even if technology and cost obstacles could be overcome, hospital administrators tended to assume that patients would resist treatment by telemedicine.[35] Another complication was that few health insurance companies and managed-care organizations were able to insure for any potential telemedicine mishaps.[36] Because these obstacles and uncertainties were impediments to profitability, hospital administrators were quite reluctant to commit themselves to telemedicine.[37] In 1995, in response to a survey asking hospital administrators to name the major barrier to adopting telemedicine technologies, 51 percent identified physician reluctance, followed by a lack of interest in and understanding of the importance of telemedicine (29 percent), and patient opposition (18 percent).[38]

Over time, the development of new technologies in the 1990s, especially advances in IT, led to improvement in telemedicine, enabling a system that was higher in quality yet smaller and less expensive. Given the technological advances and the reduced costs of telecommunications, telemedicine began to spread to a large number of patients.[39] By 1999, over twenty-five states had approved telemedicine programs.[40] (Arizona mandated commercial payment

34. Katherine Watt, "Telemedicine Extending Services Statewide," *Inside Dotcom-Tucson* (January 22, 2001): 15B.

35. T. L. Williams, C. R. May, and A. Esmail, "Limitations of Patient Satisfaction Studies in Telehealthcare: A Systematic Review of the Literature," *Telemedicine Journal and E-Health* 7 (2002).

36. Masatsugu Tsuji, Toru Mori, and Masaaki Teshima, "Beikoku ni okeru Joho Haiwei to Enkaku Kyoiku/Iryo" ["The Information Super Highway and Its Application in the Fields of Education and Medicine in the United States"], *Joho Tsushin Gakkai Nenpo* 9 (March 1998): 139–40; Mitch Gitman, "Health Via Computer," *Arizona Daily Star* (August 11, 1997).

37. "Telemedicine: Part of the IT Revolution," *Futuretech* (February 8, 2002).

38. S. Emery, "The Diffusion of Telemedicine in the Southeastern United States: A Rural-Urban Perspective," North Carolina Rural Health Research Program, University of North Carolina at Chapel Hill, 1996.

39. J. Preston, F. W. Brown, and B. Hartley, "Using Telemedicine to Improve Health Care in Distant Areas," *Hospital Community Psychiatry* vol. 43, no. 1 (January 1992): 26–27.

40. Mark J. Syms and Charles A. Syms III, "The Regular Practice of

for telemedicine reimburse for telemedicine consultations in January 1999, and in December 2000 the U.S. Congress passed legislation that increased Medicare reimbursement for telemedicine services.)[41] In 1993, only about 1,750 cases of telemedicine consultation were reported in the United States, but by 1998 the number of such consultations had reached 58,000.[42] The overall U.S. telemedicine market was estimated to be close to $2 billion per year in 2006.[43] Jonathan Linkous, executive director of the American Telemedicine Association, is convinced that the "Internet and modern telecommunications are transforming telemedicine into a powerful tool. It is changing the nature of medicine and making medical care more accessible and available."[44]

Telemedicine became politically attractive to many state governments because, in a highly visible way, it uses cutting-edge telecommunication technology to deliver medical resources normally associated with urban areas to people living in isolated rural areas without requiring a large portion of the state government budget.

In July 1996, the Arizona state legislature appropriated $1.2 million to establish the ATP. Located at the Arizona Health Sciences Center on the Tucson campus of the University of Arizona, ATP became part of the Arizona Rural Telemedicine Network (ARTN), a private asynchronous transfer mode network. The ARTN connects fifteen locations in nineteen communities across Arizona.[45] From 1996 to 2003, there were nearly 100,000 teleconsultations in Arizona. Teleradiology was the most frequently used telemedicine service, comprising over eight-five thousand cases, followed by teledermatologic and telepsychiatric consultations.[46]

Telemedicine," *Archives of Otolaryngology* vol. 127, no. 3 (March 2001): 333–34; "Telemedicine: Part of the IT Revolution."

41. "Telemedicine: Part of the IT Revolution."

42. Timothy F. Kirn, "Telemedicine Finally Making Inroads," *Family Practice News* vol. 30, no. 14 (July 15, 2000), 32.

43. "Investments Tips for Aggressive Traders! June 29, 2006," *Market Wire* (June 2006).

44. Samuel Greengard, "Telemedicine Today," *IQ Magazine* (November/December 2002).

45. Quoted in Eric J. Adams, "Technology Oasis in Tucson, Arizona," *Business Industries and Solutions* (May/June 2002).

46. Ana Maria Lopez, Deirdre Avery, Elizabeth Krupinski, Sydney Lazarus, Ronald S. Weinstein, "Increasing Access to Care via Tele-health: The Arizona Experience," *Journal of Ambulatory Care Management* vol. 28, no. 1 (January–March 2005): 16.

One of the ATP's unique features is that it offers a broader clinical scope than almost any other telemedicine program in the United States. The ATP is an amalgam of multidisciplinary specialists from several different departments and divisions at the University of Arizona Health Sciences Center. The specialists come from the fields of teleradiology, teledermatology, telepsychiatry, telecardiology, teleorthopedics, teleneurology, telerheumatology, and many others. The ATP provides patients with the opportunity to gain specialized medical treatment without traveling long distances.[47] A statewide telemedicine network consists of a collaborative community of health-care providers and state government officials providing important health-care services such as disease prevention and pediatrics. In praising the close three-way cooperation among government, industry, and academia that is at the heart of ATP, Peter Likins, president of the University of Arizona, has stated: "The program's greatest accomplishment has been creating strong ties among the University of Arizona College of Medicine, various healthcare providers, and the state legislature to achieve the state's healthcare goals."[48]

Arizona's telemedicine program also provides the financial benefit of reducing the cost of care for the state's prison population. For example, the cost of transporting inmates to distant clinics for a single visit has been cut from $415 to $140.[49] The Arizona Regional Behavioral Health Authority Network (RBHAnet), a telemedicine program for psychiatrists, provides network users with opportunities to request videoconferences, respond to videoconference invitations, and view videoconference schedules, all of which have greatly reduced travel time. As a result, mental health services are now popular in many parts of Arizona.[50] Radiology services are

47. Kade L. Twist, "The Arizona Rural Telemedicine Network Q & A with Associate Director Dr. Kevin M. McNeil," *Digital Divide Network*; Ronald S. Weinstein, "Powering the Arizona Telemedicine Program," *Health Management Technology* vol. 22, no. 6 (June 2001): 46; Kevin M. McNeill, Ronald S. Weinstein, and Theron W. Ovitt, "Project Nightingale: A Geographically Distributed, Multi-Organizational Integrated Telemedicine Network Infrastructure," Proceedings of the 13th IEEE Symposium on Computer Assisted Radiology and Surgery, 550.

48. Adams, "Technology Oasis in Tucson, Arizona."

49. "Assisted by Marconi Technology, Arizona Builds a Thriving Telemedicine Program," *Marconi Communications* (August 2003).

50. "Telemedicine Saves Millions of Dollars for Arizona," *Arizona Telemedicine*, Sunday, February 5, 2006. *http://www.telemedicine.arizona.edu/updates/page1.htm#up2* (accessed February 25, 2008).

now available around the clock in many rural communities in Arizona. An appointment for a dermatology consultation may take months in Tucson and Phoenix, but in Arizona's rural communities telemedicine has enabled such consultations to be arranged more promptly.

Because of the availability of telemedicine, people living in Arizona's rural areas may have access to the necessary medical care in their hometown. The average travel cost for treatment has dropped dramatically from $520 to $105.[51] The ATP is now considered one of the most outstanding telemedicine programs in the United States.[52] In 2001, ATP won the American Telemedicine Association President's Award for its leading role in the development of telemedicine in the United States.

The way the ATP has developed in Arizona is similar to the development that once took place in Silicon Valley, an area hosting high-tech computer-related industries in California. A chief reason Silicon Valley generated innovation and became productive lies in the establishment of close connections among universities, industry, and government. Frederick Terman, a former vice president of Stanford University, exerted effective and critical leadership in promoting this trinity.[53] In Arizona, a similar style of leadership has been displayed by Ronald S. Weinstein, the charismatic director of ATP, who has tirelessly crisscrossed the state to establish a three-way consortium involving a major university (the College of Medicine of the University of Arizona), industry, and the public sector (the federal and state governments). Fran Turisco, consultant at First Consulting Group, a well-known nationwide health-care–focused consulting firm, believes that "telemedicine allows collaboration in a way that wasn't possible in the past. It is providing benefits for patients, physicians, and healthcare organizations."[54] This tripartite consortium has provided Arizona with economic benefits and high-quality medical care.

Arizona attracts not only younger people but also senior citizens. People aged sixty or older represented 20.5 percent of the total

51. R. S. Weinstein and K. M. McNeill, "What Works: Powering the Arizona Telemedicine Program," *Health Management Technology* vol. 22, no. 6 (June 2001): 46–47.
52. Ibid., 47.
53. Yamada, "Electronics Sangyo no Ricchi no Mechanism," 38–40, 46.
54. Greengard, "Telemedicine Today."

population movement into Arizona between 1990 and 2000.[55] Southern California and Florida were once popular locations for retirement, but the quality of life in those places has declined as both living costs and crime have increased. Consequently, Arizona has become attractive to senior citizens in search of more comfortable retirement locations that offer a better quality of life at low cost. As more and more retirees move to Arizona, the state's medical community has found itself faced with a need to provide chronic care management. As a result, health-care telemedicine will play an important role in providing the state's growing elderly population with opportunities to be cared for at home.[56]

According to the Centers for Disease Control and Prevention, an agency of the U.S. Department of Health and Human Services charged with protecting the public's health and safety, chronic diseases are defined as "illnesses that are prolonged, do not resolve spontaneously, and are rarely cured completely."[57] Common chronic diseases include high blood pressure, diabetes, arthritis, and cardiovascular disease. More than ninety million Americans suffer from chronic diseases, and the cost of their medical care amounts to approximately $400 billion, accounting for more than 60 percent of overall health-care costs.[58]

Telemedicine is proving to be a profitable system by which health-care providers can manage patients with chronic diseases in their homes. Regular and frequent home-care visits would be ideal because these would enable patients (senior citizens and others with chronic problems) to have relaxing face-to-face consultations while avoiding the discomfort and inconvenience of traveling long distances to describe their medical problems. But using telemedicine, many of these problems have been addressed. Installing a small device using phone lines or fiber optics establishes a continual line of communication between health-care providers and patients. Utilizing electronic monitoring in a home-care setting, health-care providers can send daily reminders to patients with

55. Arizona Economic Base Study and Special Reports (July 2002).
56. "Telemedicine: Part of the IT Revolution."
57. "About CDC's Chronic Disease Center," Centers for Disease Control and Prevention. *http://www.cdc.gov/nccdphp/about.htm* (accessed February 25, 2008).
58. Roy L. Simpson, "The Role of IT in Caring for the Chronically Ill," *Nursing Administration Quarterly* vol. 24, no. 3 (spring 2000): 82; Jeremy J. Nobel and Julie C. Cherry, "IT Helps Manage Patients with Chronic Illness," *Health Management Technology* vol. 20, no. 11 (December 1999): 38.

chronic diseases about which medicines and foods should be used or avoided. In short, telemedicine helps patients control their eating habits, medications, and other lifestyle factors, which is essential for effective treatment of chronic health problems.[59]

Through telemedicine, health-care providers can examine a patient's daily condition, including blood pressure, glucose levels, heart rate, lung function, body temperature, pulse, and weight, and this information can be sent to the hospital for systematic processing. Telemedicine and other IT-related measures will be increasingly useful for monitoring home-care patients with chronic diseases. Moreover, effective daily monitoring through telemedicine has reduced the rate of patient readmission to the hospital.[60]

Building and expanding broadband networks in rural areas provides great economic benefits to the telecommunications corporations located in major urban centers. In Arizona, this kind of work on a rural network helps to facilitate the introduction of the skills and foster the expertise needed in a high-tech labor force and serves to generate employment.[61] In addition, approximately twelve hundred high-tech, knowledge-intensive corporations in six targeted industries—advanced materials, aerospace, biotechnology, environmental technology, information technology, and optics—employ over fifty thousand workers, thus greatly contributing to Arizona's economic growth. These corporations all have close connections with the University of Arizona. According to Duff Hearon, former chairman of the Greater Tucson Economic Council (GTEC), "The University of Arizona is at the center of Tucson's economic revitalization and is one of the region's greatest draws for high-tech companies. When you have people from the public sector, business sector, and education sector in the same room, it makes for a very powerful presentation."[62] This statement again demonstrates that Arizona has found useful ways to combine the advantages of government, industry and higher education.

59. E. W. Campion, "Can House Calls Survive?" *New England Journal of Medicine* vol. 337, no. 25 (December 18, 1997).

60. Rita Bendekovits, "Orthopedic Nursing," *Pitman* vol. 21, no. 1 (January/February 2002); Roger Fillion, "Sharing the Health Technology Expands Long-Distance Care," *Denver Post*, April 5, 1999.

61. Dorr, "Telemedicine Program."

62. Adams,"Technology Oasis in Tucson, Arizona."

Expectations and Limitations of Telemedicine in Arizona

The ATP provides telemedicine services and distance learning (continuing medical education) to rural communities in Arizona, many of which suffer from a dearth of physicians. One of the major reasons physicians tend to avoid moving to rural areas is that they become isolated from the major medical centers, which prevents them from staying abreast of the latest information on constantly advancing medical knowledge. A rural practice can also lead to a loss of professional status because of a lack of affiliation with a major medical institution. Telemedicine is an ideal means for combating these problems. The ATP began to provide local physicians and hospitals in rural areas with continuing medical education and access to formerly difficult-to-obtain specialist opinions.[63] Through the ATP network, local physicians became connected with the major medical centers in Phoenix and Tucson, an affiliation that has given them greater prestige in their own rural communities. E-Healthcare Arizona, a statewide education program managed in conjunction with Arizona state agencies, also takes advantage of the telemedicine program, offering over 500 interactive health-education programs throughout Arizona. Today, health-care providers in 14 remote areas in Arizona have easy communication access to 270 medical specialists at the University of Arizona.[64]

Health care and economic growth are closely interrelated. ATP director Weinstein predicts that advanced telemedicine will enable a greater number of industries and retirees to move to rural areas in the near future given the attractions of lower land prices, a healthy natural environment, and increasing ease of access to the cutting-edge health-care services that used to be available only to residents of major urban centers.[65]

Weinstein's prediction, however, may be too optimistic. Tele-medicine in Arizona also has its limitations. Despite its status as a high-tech state, Arizona suffers from an uneven distribution of telecommunications infrastructure. The depth of this infrastructure

63. Brian Sodoma, "Shortage of Rural Doctors Beginning to Ease," *Inside Tucson Business* vol. 9, no. 13 (June 21, 1999).

64. "University of Arizona Wins the American Telemedicine Association President's Award Using TANDBERG's Videoconferencing Solutions," *Business Wire* (June 6, 2001).

65. "Telemedicine Extending Services Statewide."

is greatest in and around the major cities of Phoenix and Tucson, but many of the rural regions trail far behind. In many respects, Arizona is sharply divided by wealth. Among the state's native Indian population, about half the Apache and a quarter of the Navajo households lack basic telephone service. As economic activity is concentrated in the two major cities of Phoenix and Tucson and their suburbs, regions outside these areas suffer from high rates of unemployment and poverty relative to the state average.[66] In contrast to the high incomes of workers in knowledge-intensive industries living in the states' modern urban areas, those living in rural areas suffer low wages, occasional indifference by the state to human rights, continuing discrimination toward African Americans, inadequate infrastructure, economic stagnation, and a general social malaise.[67] The briefing manual for the Spring 2001 Arizona Town Hall, which provided a broad overview of Arizona's economy, asked: "Why is there such disparity of wealth among Arizona's citizens?"[68] On the basis of income, Arizona exhibits an enormous disparity, the second highest in the United States, between the most affluent and the least affluent families.[69] Prior to the rapid economic development brought about by high-tech and knowledge-intensive industries, Arizona's poverty rate was higher than the national average, and its per capita income was lower. According to the U.S. Bureau of Economic Analysis, in 2004, the state ranked thirty-eighth in per-capita personal income.[70]

Another major obstacle to the adoption of telemedicine is that excessive dependence on cutting-edge technology may destroy an essential component of health care: the venerated relationship between physicians and patients. It has traditionally been assumed that a physician-patient relationship that can generate emotional

66. Arizona Economic Base Study and Special Reports (July 2002).

67. Thomas A. Lyson, *Two Sides of the Sunbelt: The Growing Divergence between the Rural and Urban South* (New York: Praeger, 1989); Atsushi Fujioka, *Sunbelt American South: Structural Outline of Polarization* (Tokyo: Aoki Shoten, 1993) (in Japanese).

68. Arizona Town Hall is an independent, private, nonprofit civic organization created in 1962 to establish, through research and discussion, an ever-increasing body of Arizona citizens accustomed to the processes of searching analysis and well informed on the many facets of the state's economic, cultural, and social life.

69. Catts, *Report of the 78th Arizona Town Hall*.

70. "2004 Per Capita Personal Income Levels and Ranks," Per Capita Income Growth Accelerated in 2004 (March 28, 2005). *http://www.bea.gov/newsreleases/regional/spi/2005/pdf/spi0305pc_fax.pdf* (accessed February 25, 2008).

comfort and compassion can only be firmly established through face-to-face encounters.[71] At a practical level, experience shows that elderly people in rural areas are less familiar and less comfortable with using telemedicine devices than are urban inhabitants, and they have a tendency to be reserved in divulging personal information using telecommunications devices.[72]

To summarize, the telemedicine program run by the ATP combines cutting-edge telecommunications and information technologies to enable delivery of health-care services to rural areas. At first glance, the ATP and the University of Arizona as a whole appear to provide Arizona with an economic stimulus and rural areas with the latest health-care services. In reality, however, disparities in wealth, income, and living standards between the greater Phoenix and Tucson areas on the one hand and rural areas on the other have become wider and wider. Meanwhile, the state enjoys a tax burden and a level of government spending that are lower than the national average. The conclusion is that while Arizona aims for small government, unfettered business competition, and a market-oriented state economy, it has little interest in providing the state's less well-off with social welfare benefits and constructing an equitable society. For the moment, this liberal market economy strategy has contributed to the expansion of the disparity between the metropolitan centers and the rural periphery even as it maintains a stable social order under the control of predominantly white technocrats and wealthy senior citizens.

Concluding Observations

For three decades, the Sunbelt region of the United States has seen the country's greatest growth, both in terms of economic growth and of political power. In recent years, Arizona has been emerging as a star performer of the New Sunbelt, effectively utilizing a government-industry-university collaboration and focusing on developing high-tech and knowledge-intensive industries. These

---

71. B. Stanberry, "Telemedicine: Barriers and Opportunities in the 21st Century," *Journal of Internal Medicine* vol. 247, no. 6 (June 2000).

72. P. Whitten, "E-Health Holds Promise and Many Questions," *Aging Today* vol. 22, no. 6 (2001): 9–11.

features are embodied in the development of telemedicine in Arizona. Despite the many obstacles to expanding the use of telemedicine, Arizona has been leading the way in the United States, spearheaded by a charismatic leader, Ronald S. Weinstein. Telemedicine has many advantages over conventional methods, and it is becoming popular in Arizona. However, the disparity between the accessibility of telemedicine services in metropolitan centers and that in the rural periphery remains; it may even be increasing. Addressing this disparity, a condition that defines the continuity between the traditional and the Sunbelt South, is a big challenge for Arizona in the twenty-first century.

Contributors

Sean Patrick Adams is Associate Professor of History at the University of Florida, where he teaches classes in the history of American capitalism, American slavery and abolition, and the Early American Republic. He is the author of *Old Dominion, Industrial Commonwealth: Coal, Politics, and Economy in Antebellum America* (2004).

Susanna Delfino is Associate Professor of History and Institutions of the Americas at the University of Genoa, Italy. She is the cofounder and currently First Vice President of the Southern Industrialization Project (SIP), which sponsors the series New Currents in the History of Southern Economy and Society. Her recent publications include *Neither Lady nor Slave: Working Women of the Old South* (2002), and *Global Perspectives on Industrial Transformation in the American South* (2005), both coedited with Michele Gillespie.

Pamela C. Edwards received her Ph.D. in History from the University of Delaware. She currently teaches at Shepherd University in Shepherdstown, West Virginia, and is working on a book manuscript titled, "Business, Technological and Labor Networks in the Development of the New South Textile Industry."

Richard Follett is Reader in American History at the University of Sussex in Brighton, England. He is author of *The Sugar Masters: Planters and Slaves in Louisiana's Cane World, 1820–1860* (2005) and numerous other essays on plantation slavery, demography, and nineteenth-century rural history. Follett served as founding editor of the Routledge journal *Atlantic Studies,* works on black expressive culture in Brazil, and is completing a project entitled "Race and Labor in the Cane Fields: Documenting Louisiana Sugar, 1845–1917."

Michele Gillespie is Associate Provost of Academic Initiatives and Kahle Associate Professor of History at Wake Forest University. Her publications include *Free Labor in an Unfree World: White Artisans in Slaveholding Georgia, 1790–1860* (2000). She has coedited two other books with Susanna Delfino, *Neither Lady nor Slave: Working Women of the Old South* (2002) and *Global Perspectives on Industrial Transformation in the American South* (2005), and has been involved with SIP since 1996.

Robert Gudmestad is an Assistant Professor of History at Colorado State University. He is the author of *A Troublesome Commerce: The Transformation of the Interstate Slave Trade* and is currently writing a book about the influence of steamboats on the antebellum South.

Yoneyuki Sugita is Associate Professor of American History at Osaka University. He received his Ph.D. in American History from the University of Wisconsin–Madison in 1999. He is the author of *Pitfall or Panacea: The Irony of U.S. Power in Occupied Japan, 1945–1952* (2003). He has published widely on health care in the United States and Japan.

Stephen Wallace Taylor received his Ph.D. in Southern History from the University of Tennessee. He is an Assistant Professor of History at Macon State College in Macon, Georgia, where he specializes in the history of industrialization, tourism, and land use. He is the author of *The New South's New Frontier: A Social History of Economic Development in Southwestern North Carolina* (2001) and has been a member of SIP since 1997.

Gavin Wright is the William Robertson Coe Professor of American Economic History at Stanford University. His publications on the economic history of the American South include *The Political Economy of the Cotton South (1978), Old South, New South* (1986) and *Slavery and American Economic Development* (2006). Wright's current research is on the economic causes and consequences of the Civil Rights Revolution.

Index

Abolitionism, 116, 117
Adams, Sean Patrick, x, 11–12, 41–67
Aetna steamboat, 24
Agricultural associations, 76, 96
Agriculture: cost of implements and
   machinery per improved acre, 89;
   mechanization of, 5–6, 73–74, 78,
   81–83, 87–92, 95–96; of Native
   Americans, 69; rice production,
   67, 68; technology transfers in, 67;
   tobacco cultivation, 33, 47, 65.
   See also Cotton production; Plan-
   tations; Sugar industry
Alabama: Dwight Manufacturing
   Company in, 126–27; foundries
   and machine shops in, 9; iron
   industry in, 9; and steamboats,
   29; textile industry in, 104,
   126–27, 147n66
Alcoa, 171, 172
Alexandria Democrat, 80
Allen, R. L., 80
Aluminum Company of America,
   171, 172
American Agriculturist, 80
American Telemedicine Association,
   193, 195
Ampère, Jean Jacques, 68, 83
Anderson, Eric, 153
Anderson, Jean Bradley, 156
Antebellum southern industrializa-
   tion: and agriculture, 5; contem-

porary comments on, vii–viii; and
slavery generally, viii, xi–xii, 8;
statistics on, viii. *See also* Coal
mining; Steamboats; Sugar industry; Textile industry; and specific
states
APNE (Arizona Partnership for the
   New Economy), 189–90
Arab oil embargo, 176
Arizona: area of, 191; cities in, 184,
   187, 188; continuing medical education in, 198; corporations
   located in, 189–90; elderly population in, 195–96; fastest growing
   major cities in, 184; house prices
   in, 188; income disparity and
   poverty in, 199; Native Americans
   in, 199; population of, 187–89,
   195–96; rural population of, 191;
   taxation in, 189, 190; telecommunications corporations in, 197;
   telemedicine in, 14–15, 181,
   190–201
Arizona Partnership for the New
   Economy (APNE), 189–90
Arizona Regional Behavioral Health
   Authority Network (RBHAnet),
   194
Arizona Rural Telemedicine Network
   (ARTN), 193
Arizona Telemedicine Program
   (ATP), 14–15, 181, 190–201